AF530988

SPACE AND TIME

BY

EMILE BOREL

With a New Foreword by

BANESH HOFFMANN
Department of Mathematics
Queens College, New York

DOVER PUBLICATIONS, INC.
Mineola, New York

Copyright

Bibliographical Note

This Dover edition, first published in 2003, is an unabridged and unaltered republication of the 1960 Dover reprint of the first English translation, originally published in 1926 by Blackie & Son, Ltd., Glasgow.

Library of Congress Cataloging-in-Publication Data

Borel, Emile, 1871–1956.
[Espace et le temps. English]
Space and time / by Emile Borel.
p. cm. — (Dover phoenix editions)
Previously published: New York: Dover Publications, 1960. With a new foreword by Banesh Hoffmann.
Originally published (as an English translation): London ; Glasgow : Blackie & Son, 1926.
Includes indexes.
ISBN 0-486-49545-0
1. Space and time. 2. Relativity (Physics) I. Title. II. Series.

QC173.59.S65B6713 2003
530.11—dc21

2003053068

Manufactured in the United States of America
Dover Publications, Inc., 31 East 2nd Street, Mineola, N.Y. 11501

Foreword to Dover Edition

Émile Borel was a mathematician of world renown, and a man of courage and incredible energy. In his lifetime he was not merely Professor at the Sorbonne, Associate Director and later Honorary Director of the École Normale, President of the French Academy of Sciences, and Director of the Institut Henri-Poincaré — which he was largely instrumental in founding. He was also, among other things, a Mayor, a Deputy, and even, briefly, Minister of the Navy. In the first world war he saw service both at the front and at the Ministry of War, and received the Croix de Guerre. Between wars he was an active worker for the cause of the League of Nations. And in his seventies, in dire times, he aided the French Resistance — for doing which he was imprisoned by the Germans and awarded the Resistance Medal by the French.

Withal he was a prolific writer. In addition to his research papers, he was author or co-author of some fifty books: on probability, on analysis, on the theory of sets, on geometry, on mechanics; advanced treatises, text books, popularizations; a famous book on chance, a book for the layman on the mathematics of the game of bridge, and even a book on aviation.

Yet of all his books, only this one on space and time has been translated into English. And as one reads this book, the fact that no others have appeared in English becomes all the more incomprehensible. For Borel was a gifted expositor, in the distinguished tradition of his mentor, Poincaré. He wrote with authority and insight — and yet with a beguiling simplicity so cunning that it seems both artless and effortless. Indeed, effortless it must truly have been, for how else could Borel have found time for all he accomplished? But artless, no. Only those who have themselves struggled to make recondite things plain can fully appreciate his skill and artistry.

The book is not intended to be a self-contained explanation of the theory of relativity. To be sure, it is *about* Einstein's theory; very much about it. But Borel does not try to explain that theory systematically in all its aspects. Significant details of the special theory of relativity are relegated to a brief mathematical appendix. And the treatment of the general theory of relativity is so abrupt and skimpy as to be unintelligible to the uninitiated — as if Borel became tired towards the end, or some more pressing matter clamored for his attention and led him to bring his book to a hasty conclusion.

Borel is primarily concerned with the general problem of space and time, and this he treats in loving detail. Such things as his eloquent insistence on the importance of the ultimate decimal, and the roving breadth of his treatment of the relationship of mathematics to physics, and of geometry to the measurement of space and time, make his book unique among popularizations. With gentle persistence he succeeds in shocking the reader into the denial of preconceived notions that is prerequisite for a proper understanding of the theory of relativity.

Written, as it was, in 1922, the book inevitably shows signs of age. For example, the discussion of the precision of time measurement — "there are chronometers which vary little more than a second a day" — reads quaintly in this era of atomic clocks that are already, for short periods, accurate to better than one part in a million million — a second in some thirty thousand years — and are fast becoming considerably more precise.

That the book is, in places, somewhat dated is both blemish and advantage. Its very datedness gives it historical interest. How many of us remember the furious controversy that raged around Einstein's theory in its early days? Here one finds vivid echoes of that controversy, echoes that have an immediacy impossible to duplicate in more modern accounts. The chapter on recent advances, added in 1926 to bring the translated edition up to date, serves now but to illustrate the evanescence that is the lot of so much of science in the making. Of particular historical interest is the account there of the ether-drift experiments of D. C. Miller, the validity of which is now generally doubted, and the brief comment on the experiments by Einstein that is somehow nostalgically revealing of Einstein himself.

On the whole, time has dealt kindly with Borel's book. Most of it is as relevant and sprightly today as when it was written, and the passing years have clothed it with a patina enhancing its charm. It is a fast-flowing narrative of subtle fascination and unobtrusive erudition, flawed perhaps towards the end, yet unmistakably the work of a master: an unusually rewarding exploration of ideas of space and time — and entwined in it an evocation of an era.

BANESH HOFFMANN

Queens College
March, 1960

Preface

The reader will not find here a didactic account of Einstein's theories. Such an account requires the use of the formulæ of mathematical physics and would be intelligible only to readers equally at home with these formulæ and with physical theories. But, failing such an account, a sort of general survey of the theories of Einstein with a description of some of their aspects seems not impracticable. As we proceed, we shall have to speak of facts and theories well known before Einstein—theories which, when we reflect a little, make his new discoveries appear less strange and paradoxical, although not less admirable.

To make a complete survey of the theories of Einstein, we should have to traverse, not only the sciences of space and time, but also mechanics and electromagnetism (including optics). We have, however, as the title of the book would suggest, dealt mainly with space and time, introducing mechanical and electromagnetic considerations only when they were indispensable.

The introduction which precedes the eight chapters of the book is a preliminary reconnaissance. It is well for us to make this first of all in order to know where we are going in the course of the somewhat slow journey that is to follow, in which we may be in danger of losing our perspective in a maze of detail.

I am under no illusion as to the merits of this plan, but I indicate briefly at the beginning of the Introduction why such an exposition, however imperfect, did not seem quite useless.

In fact, the essential points of Einstein's theory now form part of general culture, like the sphericity of the earth and its rotation round the sun. The importance of the new theory of relativity from the point of view of culture is, however, quite distinct from its practical importance. It is of no consequence to an architect that the earth is round; it is even essential for him, when he is planning a house, to treat all vertical lines as parallel, that is to say, to take the earth as flat. And yet we should be rightly shocked to find an architect ignorant of the fact that the earth is round.

The theory of relativity has encountered objections and aroused controversies. I do not deny the element of truth which some of the objections may contain, nor the utility of some of the controversies; but I do not believe that the best way to serve science is to adopt towards Einstein's theories the negative attitude which was adopted towards Maxwell's work by a learned French physicist who died prematurely a few years ago. However ingenious and sometimes apparently well-founded these criticisms may be, Maxwell's equations remain the solid foundation of electromagnetism, whilst the controversies to which they gave rise possess only an historical interest.

In my opinion, one fact overshadows all the theoretical disputes: Einstein not only has gone beyond the physics of the nineteenth century in the co-ordination of known phenomena, but has also added to this co-ordination the prediction of new phenomena, and his predictions have so far been confirmed by the test of experiment. Even if, by making use of his work, we succeeded, with the aid of some analytical device, in co-ordinating the old and the new results, while taking as starting-point the methods of nineteenth-century physics, we should hardly prove the superiority of the old methods over the new. We should only be succeeding after the event where the eminent scientists of the nineteenth century had failed. It will only be when someone, starting from principles different from those of Einstein, succeeds both in foretelling new phenomena and in co-ordinating the old ones, that the new principles he uses will either take a place beside those of Einstein, or perhaps even replace the latter entirely.

In the brief notes which are appended I have given a few developments of a more technical nature than those of the text.

The present edition includes a new chapter (Chapter VIII) in which are briefly summed up and discussed the principal theoretical and experimental studies which have appeared since the publication of the previous edition.

The book has been translated from the French by Angelo S. Rappoport, Ph.D., B. ès L., and John Dougall, M.A., D.Sc., F.R.S.E.

Contents

INTRODUCTION

CHAPTER I

GEOMETRY AND THE SHAPE OF THE EARTH

CHAPTER II

SPACE AND TIME IN ASTRONOMY

CHAPTER III

ABSTRACT GEOMETRY AND GEOGRAPHICAL MAPS

CHAPTER IV

CONTINUITY AND TOPOLOGY

CHAPTER V

THE PROPAGATION OF LIGHT

CHAPTER VI

THE SPECIAL THEORY OF RELATIVITY

CHAPTER VII

THE GENERAL THEORY OF RELATIVITY

CHAPTER VIII

RECENT THEORETICAL AND EXPERIMENTAL RESEARCHES

NOTE I

NOTE II

NOTE III

THE MATHEMATICAL CONTINUUM AND THE PHYSICAL CONTINUUM

NOTE IV

THE UNIVERSE—IS IT INFINITE?

SPACE AND TIME

INTRODUCTION

FROM NEWTON AND POINCARÉ TO EINSTEIN

Curiosity about Einstein's Theories.

Many learned people are astonished, and even a little indignant, at the curiosity universally aroused by the theories of Einstein. " Here are new theories," they say, " which we find it very difficult to understand. Only those among us who have a competent knowledge of both mathematics and physics in their most modern aspects are able to make any attempt to assimilate them, and even they do not always succeed. Unable to understand these theories, we are obliged, if not to reject them, at least to reserve our judgment as to their value, and are tempted to suspect that their importance has been somewhat exaggerated by a few enthusiasts and visionaries who have been attracted by their very strangeness. In any case, if there is anything interesting to be derived from these theories, it is a matter for the specialists. Let them work in peace. It is, however, rather strange to notice that not only philosophers, but even the general public, under the pretext that the question is simply one of space and time —and everybody thinks he knows what these mean—are manifesting a vivid curiosity about Einstein, his personality, and his theories. All this is no doubt the fault of the press, which is always ready to take up any subject so long as it promises to be sensational. Leave the scientists alone; let them continue their work in peace and do not disturb them by premature curiosity; in ten years, or in a hundred at most, they will have managed to get to the bottom of the thing, and then we shall know whether it is worth our while to take an interest in it."

Unfortunately, however, the public pays but little heed to such wise counsels; it is not in a hundred years that it wants to be enlightened and instructed, but at once. And if those who are qualified to instruct it refuse to do so, then the public will perforce adopt the explanation of these new theories given by some popularizer, whose knowledge may be derived from second- or even third-hand sources. In spite of what certain scientists are saying, the public feels that there is something here which is of interest to every cultured man; and the obstacles, far from frightening it away from the subject altogether, make it, on the contrary, only the more anxious to understand this something so strange and mysterious.

These obstacles, however, are real, and to deny their existence would be childish. In spite of the numerous expositions already published,[1] some of them with a great

[1] We cannot give a complete bibliography of works and articles on the theory of relativity, not even of those written in or translated into French. The following list is arranged as far as possible in order of publication, and with no idea of assigning relative values to works on which opinion is much divided, many of which have appeared while this book was in the press: Emile Borel, *Introduction géométrique à quelques théories physiques* (Gauthier-Villars, 1914); Lémeray, *Le principe de relativité* (id.); Einstein, *La théorie de la relativité restreinte et généralisée* (id.); Lucien Fabre, *Les théories d'Einstein* (Payot); Eddington (trad. Rossignol), *Espace, temps, gravitation* (Hermann); Lémeray, *Leçons élémentaires sur la gravitation* (Gauthier-Villars); Ch. Nordmann, *Einstein et l'Univers* (Hachette); Weyl (trad. Juvet-Leroy), *Temps, espace, matière* (Blanchard); Gaston Moch, *La relativité des phénomènes* (Flammarion).

We ought to mention also the numerous lectures given by M. Paul Langevin to various learned societies, and the notes published at the end

deal of science and much talent, many people still admit that they do not yet understand the subject, and are asking for supplementary explanations. I believe that it would be difficult to give such explanations if one professed to be going to the very bottom of the new theory, and to be bringing within the reader's view the whole of its singularly complex beauty. It is as if we were going to explain the origin of the Great War to a man who, although very intelligent, knows nothing of either history or geography and does not understand any European language; a long preliminary initiation would be required. In the same way, a vast deal of knowledge, even if only of the language and terminology of mathematics, is indispensable for anyone who is really anxious to grasp and understand Einstein's work. It is not, however, necessary to master the new theories entirely in order to guess what there is really new in them that they are offering to the human mind, just as little as it was necessary to go over the calculations of Kepler and of Newton in order to admire the beauty of the law of universal attraction. Einstein has given us not only a new theory of physics but has also taught us a new manner in which to look at the world. Henceforth, it will be impossible for those who have studied his theories to think as they would have thought had they not studied them. Of course, everybody reacts in accordance with his

of 1921 in the *Comptes rendus de l'Académie des Sciences* by M. Paul Painlevé, M. Emile Picard, and M. Paul Langevin.

EDITOR'S NOTE.—For the benefit of English readers the following works may also be mentioned: A. S. Eddington, *The Mathematical Theory of Relativity*; E. Cunningham, *Relativity, the Electron Theory, and Gravitation*; A. N. Whitehead, *The Principle of Relativity*; A. A. Robb, *A Theory of Time and Space*; H. Weyl, *Space—Time—Matter* (translation by H. L. Brose); L. Silberstein, *The Theory of Relativity*; J. Rice, *Relativity*; G. D. Birkhoff, *Relativity and Modern Physics*; T. Levi-Civita, *Absolute Differential Calculus* (Tensor Calculus; trans. by M. Long); A. S. Eddington, *Space, Time, and Gravitation*; A. Einstein, *Relativity, the Special and the General Theory* (translation by R. W. Lawson); G. B. Jeffrey, *Relativity for Physics Students*; C. P. Steinmetz, *Four Lectures on Relativity and Space*; T. P. Nunn, *Relativity and Gravitation*; B. Russell, *The A B C of Relativity*. The last six of these are more elementary than the others.

own personality towards any thought coming from the outside, and it is quite possible that the ideas inspired by a Poincaré or an Einstein would sometimes be disowned by their inspirers. That, however, is of little importance. You can conquer the world only by allowing yourself to be partially assimilated, in other words, to be distorted by the world; such has been the fate of all great thinkers, be they philosophers, scientists, or founders of religions.

The majority of minds require this preliminary assimilation or adaptation; if you try to carry them up all at once to the unexplored summit, you run the risk of making them so dizzy that they will see nothing at all.

I will try to add my modest contribution to this necessary task by briefly examining the *limits*, or *scope*, of geometry in the light of the ideas of Poincaré and Einstein. We will then study chronology, or the measurement of time, and will ultimately be led to conclude this introduction with a few remarks on the law of universal gravitation. I hope that I shall thus have prepared those readers who may have followed me to assimilate, in their turn, all that is most interesting in the new theories for those who do not wish to study science, but are only anxious to understand the general ideas which science has brought to light.

Geometry a Physical Science.

Let us first make a digression for the benefit of those readers who are acquainted with the philosophical works of Poincaré, and who may be rightly surprised at the apparent contradiction between Poincaré's conclusions and those at which we shall arrive. This contradiction is simply due to the fact that the term *geometry* is applied to two very distinct sciences.

In fact, geometry is both an experimental and an abstract science; it is with the experimental science that we are going to occupy ourselves, for the scope and limits of the abstract science have been definitely fixed by Poincaré's criticisms. The origin of his criticism is to be found in the discovery due to the genius of Descartes, who was the first to show how by employing a system of co-ordinates—to which his name is attached—every geometrical question can be reduced to an algebraic one. Geometry, as a concrete science, is thus replaced by an abstract science, that of analytical geometry, the study of equations being substituted for that of figures. We can make a further step—consider the Cartesian co-ordinates as quantities given *a priori*, and take the equations as the very definition of geometrical entities. Anyhow, it is thus that one is obliged to proceed when studying geometry of more than three dimensions, or, in space of two or three dimensions, the geometry of the so-called imaginary figures which can have no concrete but only an algebraic representation. The mathematicians of the nineteenth century had thus gradually come to consider geometry as an *ensemble* of algebraic and analytical formulæ, an ensemble of formulæ which it is particularly interesting to study, on account of the facility with which geometrical language enables us to express briefly certain properties which algebraically are somewhat complicated. Geometry proper is thus reduced to a *schema*, and there is no sense whatever in asking whether this schema be true or false. Algebraic formulæ can be interpreted by a great number of other schemas, differing among themselves only in the degree of convenience they offer. On the other hand, the *ensemble* of these formulæ possesses that degree of absolute truth which is common only to all pure constructions of the human mind; they are as true as the statement that two and two make four. In this algebraic conception the question of the limits of geometry does not even arise, there being no limits to the indefinite development of formulæ; and the truth of these formulæ admits of no limitation, for it partakes of the absolute character of arithmetical truths.

The possibility of reducing geometry to a purely abstract analytical and algebraic theory must not, however, make us lose sight of the concrete origin of geometrical *concepts*. When Hilbert tells us to think of three systems of *things* which we are to call points, straight lines, and planes, these *things* possessing by definition such properties as that *between two points we can draw a straight line and only one*, we know very well that Hilbert would never have thought of those *things* had not Euclid lived before him.

It is with geometry as a physical science that we are going to deal exclusively. From this point of view, we may say *en passant*, the much-discussed question with regard to the number of dimensions of space is quite simple: space is *three-dimensional*, because volumes are proportional to the *cubes* of lengths. This remark, however, being outside our subject, we will close our digression.

Invariable Bodies and Various Scales.

The origin of geometry is usually traced to the endeavour of the ancient Egyptians to reconstitute the boundaries of their fields after the rise of the Nile. The methods they were led to employ did not differ very essentially from those which, with the help of a cadastral survey, will make it possible for us to discover and restore the boundaries of rural and urban estates in the devastated regions of France and Belgium. To make such a reconstitution possible, it is above all a *sine qua non* that the cataclysm, be it the rise of the river or the invasion of the enemy, should not have modified the dimensions of the earth. In its historical origin geometry is thus based upon the postulate of the existence of invariable solid bodies. It is a postulate which, on account of our being so familiar with it, we are often tempted to forget; it is not only our geometry, but also our entire daily life which presupposes the existence of invariable landmarks, such as our house, our fields, and our streets. A real effort is required if we wish to imagine the ideas of a fish which lives constantly in the ocean and has never perceived either the shores or the bottom of the sea. Supposing such a fish to be endowed with intelligence and senses analogous to ours, it would never perceive anything but the surface of the water agitated by waves, and other fishes in continual motion with respect to itself. There would be no fixed landmark which could serve as a support for a geometrical construction. We are not going to waste our time in discussing the information such a fish could derive from the contemplation of its own body or other bodies analogous to its own. The difficulties would certainly be great, especially if the fish happened to be carried away by currents, of the more or less complicated nature of which it is ignorant.

Thus the first condition necessary for the creation of geometry is the existence of objects which to our eyes are immovable, that is to say objects which remain *sensibly* invariable when we consider them *on our own scale.* A few words of explanation will not be superfluous with regard to the double fact, namely that this immobility is only approximate, and that it is observable only in objects on our own scale. I am seated at my desk and looking from time to time at the furniture and the walls of my room; they appear to me to be motionless, and yet a more attentive observation and closer examination would enable me to notice sometimes a shaking and tottering of the building, produced perhaps by a passing omnibus or train in the neighbourhood, more rarely by a shock of the crust of the earth, a shock which a responsive seismograph would not fail to register. But this is not all; temperature is not strictly constant, and I know that pieces of various metals expand in an unequal degree; wood dries up and often emits a crackling noise. If I am therefore anxious to make precise observations and avoid these small perturbations, I shall have to fit up a laboratory with concrete pillars isolated from the walls, and directly supported by a deep and solid foundation of rock. I should also have to keep up an invariable temperature. I do not, however, insist upon these difficulties, insurmountable though they be, and merely *en passant* will I call attention to the fact that the earth, though it appears to us fixed and stationary, is not motionless in space. I manage, however, after a fashion, to define bodies which are sensibly stable in relation to each other, and I can measure their respective distances—this is the beginning of geometry. Next, I observe that this result is obtained only for objects on my own scale, or at least, for objects the dimensions of which are neither too big nor too small in relation to mine. The biggest solid object the dimensions of which I may ever hope to ascertain by means of direct measurements is the earth, the observation of the stars requiring such a complicated mass of hypotheses pertaining to physics that it can no longer be looked upon as purely geometrical. Besides, the stars are in perpetual motion in relation to each other, and, on their scale, we can see no fixed and stable landmark in the universe to which we could refer their movements. As for the landmarks which we are able to imagine, such as the centre of gravity of the solar system, these are abstract conceptions whose definition requires a knowledge both of physics and of mechanics which will lead us far away from geometry proper. At the other extremity of the scale, again, in the infinitely small, we know that the atoms are in a state of perpetual vibration, and that a being living on their scale, like a being living on the scale of the stars, would no longer be able to observe even that relative immobility which sufficed to enable us to constitute our geometry.

Thus, as soon as we diverge to any extent from the human scale, two limitations to this geometry become immediately apparent, one in the very great, the other in the very small. It is remarkable enough that we find ourselves almost exactly in the middle between these two limits; the dimensions of the earth are about ten million times as big as those of man, whilst the dimensions of an atom are about the ten-thousand-millionth part of ours; it is between these limits that our geometry exists. In choosing the *metre* as a standard of measure, we have chosen a standard which is at once a standard on our own scale and sufficiently accurately intermediate between the biggest and the smallest lengths which we can measure directly (that is to say without having recourse to optical and astronomical operations). We know that the originators of the metrical system chose the quarter of the terrestrial meridian as the fundamental unit of length, the metre being defined as the ten-millionth part of this quarter of the meridian.

The fundamental unit of length was thus the greatest length directly accessible to us. It has since been frequently suggested to take the wave-length of a fixed radiation as a unit of length, that is to say one of the smallest lengths we can reach. The choice, however, has been fixed on the metre, that is upon a certain international standard originally defined by means of the terrestrial meridian, which, however, in view of the growing precision of geodetical measurements, we would constantly have to modify, were we to retain the original definition. The metre has anyhow been very accurately measured in wave-lengths, the smallest standard thus being brought into relation to the normal one, and there is no doubt that we shall also in time reach very high accuracy in measuring the terrestrial meridian in metres. Indeed, the metre is the standard of measure on our scale which is actually adopted as the most precise.

But even this standard cannot be defined with an error inferior to molecular dimensions, and this is another limit to the precision of all experimental geometry.

Geometry inseparable from Optics.

But is there any experimental geometry? Everybody agrees that it would be a waste of time to try to verify experimentally the accuracy of geometrical relations, unless we were to consider this verification as an experiment in *physics*.

In other words, if, after having measured with accuracy the sides and angles of a triangle, we found that the figures obtained did not strictly agree with trigonometrical formulæ, we would look for the explanation of this small difference in a physical phenomenon and not in geometry. We should ask ourselves, for instance, whether the ruler with which the measurements have been made has not expanded in consequence of heat or some other physical factor, all experiments hitherto made being in agreement with the hypothesis that geometrical relations are exact, and that the error, if it does exist, is inferior to experimental errors. We are therefore naturally entitled to think that such will be the case when the accuracy and precision of the measurements is increased. Let us admit this assumption so as not to complicate the discussion. We shall, however, run up against the limits resulting from the greatest and the smallest dimensions of the permanent solids we are able to compass (the terrestrial globe and the atoms); above and below these there would be no sense in imagining that the experimental verification would be satisfactory, since we have no means of conceiving such a verification. Astronomers, however, estimate the distance of the stars, whilst physicists measure with very great precision the wave-lengths of luminous vibrations. In both cases recourse is had to the peculiar properties of light, and we cannot try to pass the limits of geometry unless we make use of these properties. We shall thus no longer be studying pure geometry but both geometry and optics.

It is remarkable that we should be able to make use of the peculiar properties of light for the purpose of estimating both very great and very small lengths which are beyond the reach of the geometry of solid bodies. Light travels 300,000 km. in a second, that is to say in the shortest space of time accessible to our senses it travels a distance equal to the dimensions of the earth, whilst, on the other hand, luminous wave-lengths are of the order of one-thousandth of a millimetre; the human scale is almost in the middle between these two scales furnished by optics; it is a fact deserving consideration. But let us pass on. What is most important for us at the present moment is the fact that it is impossible to separate geometry from optics. This impossibility, now that our attention has been called to it, will appear to us even within measurements on the human scale, if we are anxious to arrive at very great precision. We have indeed noticed that absolute immobility is merely apparent. The mechanical theory of heat teaches us that absolute immobility could not have any chance of realization except at absolute zero, that is to say 273° below the zero of the centigrade thermometer; this would, of course, be of no earthly interest to us, as we should long be dead by then. Life is incompatible with rest. It is therefore impossible to speak strictly of geometrically invariable figures,[1] and we shall find ourselves compelled to study geometry at a given instant, then again at another, and so on, so as to gain a sort of cinematographic image of the world. This is a new restriction of geometry which deserves to be examined separately.

Difficulties due to Motion.

Let us now suppose that, after having perfected to an even greater extent the most sensitive cinematograph, thanks to which it has been found possible to study the

[1] We may point out in passing the value of the study of mean positions, as defined by the methods of the calculus of probabilities.

motion of projectiles, we succeeded in obtaining absolutely instantaneous photographs, the time of exposure being reduced to the one-thousand-millionth of a second. Shall we then have an exact picture of the objects surrounding us at the moment the photograph was taken? It would be so, were the propagation of light absolutely instantaneous and its travelling speed mathematically infinite. We know, however, that such is not the case, and that light travels approximately a distance of three hundred million metres in a second. Given now an object situated at a distance of three metres from the objective of our photographic apparatus, it follows that the one-hundred-millionth part of a second has pased between the moment when the photograph was taken and the moment when the luminous ray left the object. What we are really fixing on the photograph is no longer the position of the object at the very moment of photographing, but its position a hundred-millionth part of a second earlier. Supposing next the object to be situated at a distance of thirty metres from the objective, we would be fixing its position a ten-millionth part of a second earlier. The geometrical figure photographed does not at any instant therefore correspond to reality, since the two objects have in the intervening moment been displaced. If they have been displaced as rapidly as projectiles, if their speed attains, for instance, one thousand metres per second, then it follows that in the ten-millionth part of a second they will have been displaced to the extent of a tenth part of a millimetre, a distance which cannot be considered as entirely negligible. Thus, in spite of the most perfected means of investigation imaginable,[1] the study of the geometry of bodies

[1] I leave out of account one conceivable method, which consists in moving about a solid body of sufficient size so as to bring two fixed points upon it into coincidence at the same moment with two variable points which we are studying. In point of fact a solid body cannot be moved about without undergoing elastic deformations, the velocity of transmission of which is very much below that of light.

in motion at a given moment (and all bodies are more or less in a state of motion) cannot be separated from or undertaken independently of a study of the velocity of light.

Other difficulties, however, then arise, difficulties to which for the present it will suffice to call attention without our attempting to solve them. Let us imagine a circular platform or floor, something like a roundabout with wooden horses, which may remain motionless or be set in motion in the interior of a fixed circus. Let us say, in passing, that the earth upon which we live resembles, in a certain measure, such a platform, with the only difference that the fixed circus is not accessible to us, as we can leave the platform only in imagination. Now, taking up our position upon the movable platform, we can mark a point A and several other points B, placing at every point thus marked a well-regulated clock, in such a way that if we give a signal at the point A we shall perceive it at each of the points B at the moment at which the signal was given *plus* the time required for the light to travel the distance between the two points. It will be easy for us then, if we locate at one of the points B a cinematographic objective, to make the correction necessitated by the velocity of light. We can also do the same with regard to the fixed circus in the interior of which the moving platform is supposed to be situated; in this circus, too, we can have fixed points, with measured distances, and clocks placed at these points. The difficulty that will now arise will be as follows: on the edge of the moving platform we have marked two points, A and B, which, when the platform is at rest, are facing exactly the two points A′ and B′ of the fixed circus. Indeed, A′ coincides with A, and B′ coincides with B. Now we are setting our platform into rapid motion, and when A and A′ are again coinciding we send a luminous signal at this point; it takes a certain time for this signal to reach B (or B′), and when it has reached it, the two points B and B′ no longer coincide, the platform having moved in the meantime. B is therefore no longer facing B′, but another point in the fixed circus, say C′, whose distance from A′ is not at all the same as that of B′, and where the signal arrived either earlier or later than at B′. In consequence of the motion of the platform, the fixed and the movable clocks will consequently be at variance. I have no intention to go to the bottom of this question in this introduction and to examine the various solutions that have been suggested. What is important for us is to have shown that it is not so easy as would appear at first sight to define with absolute precision the geometrical position of the universe at *a given moment.* In order to be able to define this instant at different points, we should be compelled to have recourse to the properties of the velocity of light in bodies in motion with respect to one another, an operation constituting one of the most difficult problems of experimental physics. Suffice it to say that the errors which will thus arise are of the same order of magnitude as those which we have already discussed. Here, again, we find a restriction imposed upon geometry towards the eighth decimal.

Scientific Importance of an Extra Decimal.

Well, people will say, this is rather reassuring. We are certain of the exactness of seven or eight decimals; what more do we require? To go beyond this—would that not mean splitting hairs and wasting one's time upon trifles, the practical interest of which is almost nil? Such an objection will, of course, have no weight with a true scientist; the search for truth is the most noble aim of science, and there are no degrees between truth and error; the insignificance and smallness of a phenomenon will perhaps diminish its practical interest, but not its scientific value. Not satisfied with this general reply, we might go further and demonstrate by actual examples the enormous practical importance of certain observations made upon quantities apparently quite insignificant, as in the case of a great number of astronomical discoveries. The difference between elliptical and circular orbits in planetary motion, to quote one of the most famous cases, necessitated in the days of Kepler observations of extreme accuracy, considering the limited means then available; but the laws of elliptical motion made it possible for Newton to discover the law of universal attraction. More recently, the disturbances of Uranus, which led Le Verrier to the discovery of Neptune, and the secular variation of the perihelion of Mercury, which confirmed Einstein's new theory with regard to gravitation, are among the phenomena the discovery of which involved the most refined observations. We could quote many other examples. It has been well said that every important advance in the knowledge of nature has been marked by the conquest of an additional correct decimal.[1] It will require the combined efforts of mathematicians, physicists, and astronomers to enable us to push back a little the limits of geometry. We shall be compelled to put a number of questions to which, however, we can only call attention. We shall, for instance, have to ask ourselves whether universal gravitation does not modify the length of solid bodies in a manner analogous to that by which bodies are expanded by heat. But we shall return later to these questions. For the present, my only aim in this introduction is to raise, with regard to geometry, in the minds of my readers that doubt and curiosity which constitute the origin of all philosophical thought.

[1] It should be observed, however, that the conquest of an extra decimal ought not to make us forget the usefulness of earlier approximations. If, theoretically, there is an absolute gap between a very minute error and no error at all, this is not so practically—a point which might be worth special consideration.

Time a Measurable Quantity.

Let us now for a moment drop the criticism of the concept of space and try to make, in the same spirit, a few remarks on the measurement of time.

For a long time metaphysicians have been in the habit of discussing the question of the nature of time, and these discussions are not yet closed. Without intending to dispute the interest of such discussions, we, nevertheless, exclude them as being outside our province; for us, time will simply be what it is for all men, namely, an uninterrupted series of successive moments, each of which corresponds to a number which is its date. When we say that the eclipse of the sun took place on 1st October, 1921, at 10.27, or that the armistice began on 11th November, 1918, at 11 o'clock, this language, in consequence of precise conventions, conveys the same meaning to all civilized men. This division of time into years, months, days, and hours, a division consecrated by long practice, corresponds to a numerical system which is in reality rather complicated, and appears to us simple only on account of long-established habit. Theoretically, it would be more simple to estimate time in years and decimal fractions of years (or in days and decimal fractions of days). One decimal number, such as 1921·43864275, would enable us to fix a certain moment in relation to the beginning of the Christian era. This, however, is a detail of small importance; what we are anxious to prove is that not only for scientists but for everybody else as well, time is a measureable quantity, that is to say, that it is subject to number. Numbers are the landmarks which enable us to speak, in a language common to all men, of successive moments of duration.

If we wish to determine these numbers, the first thing to do will be to choose and fix in an immutable manner a time *unit*, just as we have chosen a unit (the metre) by which we measure lengths. Now how are we to choose this time unit? How are we going to avail ourselves of it? These are some of the questions which we are going to study first.

Analogy between the Measurement of Time and the Measurement of Length.

If we wish to measure lengths we utilize a ruler; and if we have to apply a metre stick three times end to end in order to find the length of a table, we say that the table measures three metres. It seems difficult, at the first glance, to proceed in the same manner with regard to time; we cannot displace an interval of time in order to verify, by means of juxtaposition, whether it equals another interval of time; nor can we place contiguously two intervals of time and thus obtain the sum total, unless these time intervals happen to be naturally juxtaposited, which is the case when the end of one coincides with the beginning of the other. How then can we measure time? We shall make use of clocks, which are either natural, such as the gigantic clock formed by the earth and the stars by means of which we count our days, or clocks carefully constructed by the hand of man, such as a simple pocket chronometer. Let us take the chronometer first, returning in a moment to the natural clocks. If a chronometer is well regulated it will indicate the time measure with extreme accuracy; there are chronometers which vary little more than a second per day, that is to say the error is about one hundred-thousandth part of the interval of time measured. It is an error which is quite of the same order as the expansion of a metallic ruler for a variation of temperature to the extent of 1° centigrade. The length of the ruler does not remain invariable when it is being displaced in space, just as the movement of the chronometer varies when it is being displaced in time. It is, therefore, as legitimate to measure an interval of time by means of a chronometer, as it is legitimate to measure a given length with a ruler.

There is, however, a great difference between the measurement of time and the measurement of lengths; we can never find again an interval of time that has passed, whilst it is quite easy to find a length and to begin over again the operation of measuring it more carefully and more accurately. This difference, however, is only apparent, for we never find again the *same* length, which has been displaced by the motion of the stars and put out of shape by the molecular motion which never ceases. It is, therefore, only the *approximate* length which we find again, and in an analogous manner we may say that *approximately* we find again the same interval of time. This is what happens when, in the course of several succeeding nights, an astronomer measures the time separating the passage of the meridian by two fixed stars. He finds that this interval of time is the same, just as we find that the dimensions of a solid body are the same to-day as they were yesterday. We know very well that the identity cannot be absolute, but the equality is very near, and it is sufficient for the requirements of our science. It is indeed because we have found that several chronometers sensibly agree among themselves, and almost strictly agree with astronomical observations, that we have come to look upon them as instruments furnishing a sufficiently exact measure of time. In the same way, we had already acquired the notion of the invariability of solid bodies, because an experiment repeated a hundred times has taught us that the dimensions of solid bodies which were equal yesterday are also equal to-day. It will be necessary to narrow down a little more this somewhat empirical definition of the duration of time and to investigate a little the mechanism of the various clocks which we employ.

Artificial Clocks and the Clocks of Astronomers.

The fundamental standards of measure proposed for the measurement of lengths may be classified in three categories: measures on the human scale, such as the metre of the Pavillon de Breteuil; measures on the scale of the terrestrial globe, such as the length of the meridian, suggested by the originators of the metrical system; and finally measures on a scale which may be termed molecular, such as the wave-lengths of light vibrations (which are considerably greater than the dimensions of molecules). The various clocks which have either been suggested or are really being utilized may also be classified in an analogous manner, with the only difference that in the case of the clocks it is those that are on an astronomical scale that have first been utilized by man. Long before any clock on a human scale had been constructed, before the sand-glass or the clepsydra, men had already acquired the habit of measuring time by days, that is to say, they had been making use of that immense natural clock, the sun. We should have to write the entire history of astronomy before we could show how the notion of the solar day, vague at first, became more clear and precise and gradually led to an exact definition of the mean solar day, the 86,400th part of which is a second, a unit of time universally recognized and adopted.

The sidereal day, shorter by four minutes than the mean solar day, is the interval of time separating two consecutive passages of the same fixed star at the meridian of the *same place*. This definition presupposes the existence of fixed stars, that is to say of stars which, with respect to each other, conserve the same relative positions. As a matter of fact, astronomical observations have shown that certain stars, originally considered as fixed, undergo slight displacements, and it is highly probable that such may also be the case with all stars. But certain stars are too far distant from us, so that it is absolutely impossible for us, with the means of investigation at our disposal, to discern their displacements. To us they appear as if they were absolutely fixed. On the other hand, these stars are too far removed from us for their course to be modified either by the displacement of our planet, the earth, in its orbital motion, or by the displacement of the solar system. Everything happens, therefore, as if the earth were motionless in space and a vast clock, constituted by the starry vault above us, were rotating round us with perfect regularity.

But even if we leave aside the phenomenon of aberration, of which we shall speak in Chapter V, and neglect the very considerable phenomenon of the precession of the equinoxes, thanks to which the axis of the world is displaced among the stars, so that the actual pole star will no longer be the pole star in a few thousand years, it is quite impossible not to insist upon the fundamental hypothesis of the strict constancy of the sidereal day. This hypothesis reduces to the mechanical hypothesis of the strict constancy of the angular velocity with which the earth rotates round its axis. Now it is clear that the latter hypothesis can only be approximate. To mention only the most apparent and best known phenomenon, the tides play the part of a brake upon this immense fly-wheel which is the terrestrial globe. We talk sometimes, not without good reason, of utilizing the energy of the tides. Here we have a source of useful energy practically inexhaustible for man. This energy cannot be created *ex nihilo*; it is in reality an infinitesimal fraction of the energy with which the earth rotates that is being placed at our disposal by the phenomenon of the tides. The sidereal day has thus a tendency to grow longer; are we able to measure the amount of this lengthening? The question may well appear insoluble, when we consider that man has as yet been unable to construct a clock approaching in accuracy the gigantic clock by means of which the sidereal day has been determined.

We can say in fact that the astronomers' clock is the whole of celestial mechanics; time is provisionally determined by means of the sidereal day, and the determination is sufficiently accurate to enable us to establish the laws of celestial mechanics, that is to say the exact explanation of observed phenomena by means of a very small number of simple principles (principles of mechanics and the hypothesis of universal attraction). This explanation is then taken as a starting-point, enabling us, if necessary, to make very slight corrections in the constancy of the sidereal day, indispensable if the explanation of the world is to remain simple. We see that we are far away from the original simplicity of the natural clock by means of which we obtained the sidereal day, this clock being replaced by an ensemble of highly complicated hypotheses and calculations. Can we hope to avoid these complications by making use of clocks on the human scale or on the molecular scale?[1] Perhaps the discussion at least of the mechanism of these clocks will be theoretically more simple.

The Influence of Gravitation upon Clocks.

The regularity of movement in clocks of precision is obtained by means of a rod oscillating with great regularity, owing to the action of gravity; this rod is called a pendulum, and we say that the pendulum beats seconds when the duration of each of its oscillations is exactly one second. This period of time depends upon two elements: the length of the pendulum and the intensity of gravity, which is not identical at all points of our planet. Pendulum observations, in fact, have furnished us with the most precise methods enabling us to study the variations in the intensity of

[1] I am disregarding celestial clocks outside the earth, such as that furnished by the eclipses of the satellites of Jupiter; to the difficulties already pointed out there is added in this case that arising from the velocity of light and the variability of the clock's distance.

gravity at different points of the earth. We can, however, avoid these difficulties by agreeing to station ourselves at a well defined spot on the surface of the earth. The unit of time could then be determined by means of the unit of length; we have only to fix with great precision the length of the pendulum, the period of oscillation of which is to be our unit of time. We may add that if we fix this length by taking as unit one of the dimensions of the terrestrial globe and place ourselves at the pole, so that the rotation of the earth can exercise no influence, then the unit of time thus determined will not depend upon the absolute dimensions of the earth, but solely upon its mean density. Thus on a globe with radius half as great, a pendulum of half the length would oscillate in the same time. On this new globe the metre would have been defined as the ten-millionth part of a quarter of the meridian, so that the pendulum of one metre in length would give us the same unit of time as upon our planet. All this is still very rough and would require a good deal of commentary and elaboration, but we are, nevertheless, able to conceive how the properties of matter may thus enable us to determine an absolute unit of time by means of a material sphere isolated in space and consisting of some pure substance accurately determined. We shall, however, have to take into account temperature, and the chemical description of this pure substance; we are thus brought face to face with thermodynamics and chemistry. Let us not, however, go out of our way, but rather content ourselves with the above definition as one furnishing us with a certain approximation. So far, this is a result of a sort. The approximation, however, is much rougher than that which we are able to obtain by means of the clocks on the astronomical or molecular scales.[1]

Let us now examine those infinitely small clocks con-

[1] For space, on the contrary, the better standard is the standard on the human scale.

stituted by the vibrations of light. We know that the duration of these vibrations is not the same for different colours, that is to say for the various parts of the solar spectrum; but we also know that in such a spectrum the physicist perceives very fine rays whose respective places seem to be accurately fixed. These rays are characteristic of diverse chemical substances, but the purity of the substances is not an absolute condition for the rays to appear. It is quite sufficient that a flame contain a trace of radium, or of iron, for the spectrum of the flame to show rays characteristic of radium or of iron. We can thus safely speak of the period of vibration of a certain ray, and are thus able to determine a unit of time, very short, no doubt, on the human scale, but still one that may be turned to account, thanks to the resources placed at our disposal by the methods of physical optics.

We shall not discuss the influence exercised upon the rays of the spectrum by electric and magnetic phenomena, for we would thus be compelled to approach one of the most interesting chapters of modern physics and to speak of the phenomenon discovered by Zeeman; this would lead us too far.

It is, however, impossible to pass over in silence the phenomenon which was predicted by Einstein and which the most recent observations of the sun made by M. Pérot seem to confirm. The theory is also confirmed by a precise analysis of observations which MM. Ch. Fabry and H. Busson made before Einstein published his theory, and which consequently cannot have been influenced by any preconceived idea.[1] According to Einstein the potential of gravitation, whose derivatives determine the field, that is to say, universal attraction, to speak in the language of Newton, exercises an influence upon luminous vibrations,

[1] See the note: ‘ Sur le déplacement des raies solaires sous l'action du champ de gravitation " (*Comptes rendus de l'Académie des Sciences*, Vol. CLXXII, p. 1020, 25th April, 1921).

which constitute the minute clock we are now considering. Gravitation produces the motions of the heavenly bodies, and it is through these motions in fact that it has been discovered. Gravitation thus directly intervenes in that gigantic clock constituted for us by the system of Jupiter and its satellites, or to take a simpler case, by the moon and its phases. Less directly, it also intervenes in the rotary motion of our planet, the earth, round its own axis. We have nevertheless found that a knowledge of all astronomical phenomena is required for an absolutely accurate astronomical determination of time, and that consequently a profound study of gravitation is implied even with regard to clocks on a human scale [1]; and now we find that gravitation intervenes also in the clocks on a molecular scale. We have here a very important fact, for it involves as a consequence that the internal motion of clocks, that is to say the absolute determination of time, is thus being modified by their external motion, or, more exactly, by the variations in the state of that movement, namely, the accelerations. We have only to enter a lift ascending rapidly to realize that at the moment of its starting or of its stoppage something like an increase or a diminution of our weight occurs; the power of attraction of the earth seems either to be increased or to be lessened. A profounder study will show us that we have no means whatever by which we can distinguish the *natural* field of gravitation, due to terrestrial attraction, from the field which we are tempted to term artificial, due to the setting in rapid motion or sudden stoppage of a lift.

The effect of the action of universal gravitation upon these natural clocks, luminous vibrations, therefore implies that the running of these clocks will be equally modified

[1] I leave out of account pocket chronometers, in which the oscillator is a small wheel connected to a hairspring; the theory of the elasticity of springs is not yet sufficiently advanced to make it possible, except empirically and by means of copies, to make a spring the balance wheel attached to which will beat seconds.

when they are subject to accelerations, that is to say, when their speed is either increased or diminished. This principle involves some very important consequences, on which it will be necessary to dwell for a little.

The Slowing Down of Clocks in Accelerated Motion.

We have tried, by various means, to determine an absolute unit of time which shall be independent of the particular conditions under which we are placed. One might have thought that the luminous vibrations would have given us such an absolute unit. But if these vibrations are influenced by gravitation and by acceleration it becomes impossible to establish any agreement with regard to the measurement of time between two observers who are stationed upon different stars, and who, in consequence of the motion of the stars, may find themselves side by side at certain intervals of time.[1] For this to happen, one at least of the observers must have had his motion modified, must have undergone certain accelerations during the interval of time between two encounters; these accelerations will have modified the movement of his clock, and clocks on the two stars, if in agreement at the first encounter, will no longer be so at the second encounter (unless, of course, the two observers have undergone equivalent accelerations, as would happen in the case when their movements are exactly symmetrical). This result, deduced by Einstein from his theories and subsequently borne out by observation of the solar rays, may be taken as our starting-point, independently of the theory which furnished it, if we consider it as a fact learned by experience. We shall, of course, have to grant to Einstein that this fact, verified in the

[1] This theoretical case is the only one which is really interesting. If two observers who happened to meet once were never to see each other again, their relative motion being such as to separate them indefinitely, the question of the agreement of their watches would not arise, since there would be no opportunity of comparing their behaviour. The only question they could consider would be a more complicated one, involving the properties of light signals.

particular case of natural clocks, such as luminous vibrations, can also be verified in the case of any other natural clock we might imagine. Generally speaking, the presence of a field of gravitation, created either by the Newtonian attraction of a big star, or by rapid acceleration, *involves as a consequence the retardation of all clocks.* Once we admit this fact, we are led to very curious consequences, demonstrated conclusively for the first time, if I am not mistaken, by M. Paul Langevin.

The universal retardation of clocks inevitably involves as a consequence the slowing down of all phenomena, for every phenomenon is, more or less, a rough clock. It must, in particular, involve the retardation of vital phenomena, and consequently the modification of what we may term psychological time, that intimate notion we have of duration, a notion which is evidently closely related to the internal phenomena of our own body. It follows that if we could manage to remain in a very intense field of gravitation and live there, *we would live less quickly*, provided, of course, that the conditions of our life were not also modified, on account of the excessive intensity of this field of gravitation. We would not advise anybody to try the experiment by submitting himself to artificial fields of gravitation which may be produced by various means, such as centrifugation, for instance. The paradoxical consequence is therefore purely theoretical, in the sense that the fields of gravitation whose intensity is compatible with our life are too weak for an appreciable effect to be noticed upon our clocks and *a fortiori* upon psychological time. Such an effect can be perceived only by means of very delicate experiments in physics, and it is only for the purpose of description that we may exaggerate and try to imagine some observer who has been set in such rapid motion that he is able to go and visit Sirius and come back to us in a century. There is, however, no doubt that if such an hypothetical traveller were to measure time by a portable clock constituted of a luminous vibrator, he would measure a total duration of time much shorter than the duration measured *during the interval between the same two events* by an observer upon the earth.

The measurement of time cannot therefore be separated from the study of universal gravitation and of motion.

The Timing of Clocks.

Hitherto we have spoken only of the measurement of duration or measurement of the regular movement of clocks, but not of the timing of such clocks, that is to say of the agreement of the noon of two clocks placed at different points. If it were possible to determine theoretically an absolute clock with a movement which, in spite of all displacements, remains absolutely regular, it would be easy, theoretically at least, to establish agreement between all the clocks in the world. We would only have to go from one point of the universe to another, carrying the standard clock with us. This, however, we do not possess, and we shall therefore find it hopeless to try to establish agreement between clocks placed at different points except by means of luminous signals. In this case, however, new difficulties will at once arise, difficulties which we have already referred to [1] and of which we shall speak again in Chapter VI.

The Necessity of Successive Approximations.

To sum up, we find that the difficulties of measuring time are as great as—if not greater than—the difficulties of measuring space; chronology, in the absolute sense of the word, is as difficult as geometry. Thus it appears that we have to give up the hope of building up the science of mechanics upon that solid basis which during the eighteenth and nineteenth centuries was thought to have been found, that is to say upon a geometry and a chronology given *a priori*. We shall have to proceed in an opposite

[1] See p. 13.

direction, and follow a method much more difficult, but also much more fertile once we have mastered it. We shall have to make geometry and chronology dependent upon the ensemble of physical phenomena, since the standards of both space and time depend upon these phenomena. To measure space and time we shall have to begin by reconstructing the whole of physics; and we thus find ourselves at the outset faced by a difficulty, since apparently we cannot even begin to make observations of physical phenomena if we do not know how to measure either space or time. This difficulty, great as it is, is in no way new in the history of science; was it not by successive approximations that we arrived at our knowledge of nature? The principal thing is that these approximations should converge. The classical methods of measuring time, methods utilized by the astronomers, and constituting one of the finest achievements of the human mind, need not therefore be entirely abandoned; they furnish us with a starting-point whence the physicists may boldly launch out in quest of new truth.

The Origin of Newton's Law.

We have thus rapidly shown how the theories of Einstein modify the notions of space and time, and, to conclude our introduction, we have only to show in what way they also modify our ideas with regard to that old and famous law of universal attraction which Newton formulated in 1682. For the first time in two centuries a step forward has been made in our knowledge concerning this law which has remained for a long time mysterious and, as it were, impenetrable, although its usual manifestation, the weight of bodies, is one of the phenomena most familiar to us. It will not be superfluous, if we wish to realize the importance of the evolution of this Newtonian law, to sketch, as briefly as possible, the most salient and essential features of its historical development.

We are told that whilst he was plunged in his deep meditations and looking up at the moon, Newton saw an apple fall to the ground. He at once asked himself: Why does the moon not fall as the apple does? Whether this anecdote be literally true or false is of no importance; it indicates happily one of the most striking features of Newton's discovery, the fact that it had its origin in an apparently trivial observation which anybody might have made. If, however, we go to the bottom of the matter, it means that Newton asked himself whether the phenomenon so well known to us as the fall of heavy bodies must be accepted as a sort of postulate unrelated to anything else, or, on the contrary, may be considered as a particular case of a more general law. To have simply asked the question already constituted the essential part of the discovery of universal gravitation, not the only great discovery which has been made simply by looking with new eyes at a phenomenon so familiar and trite that nobody has ever taken any notice of it.

The detailed account of Newton's discovery is, however, even more instructive. Let us briefly recapitulate the most salient stages. It was in 1666 that Newton began to make the calculations which, in his opinion, would enable him to find out the identity between the weight which causes the apple to fall and the attraction regulating the course of the moon in its orbit. Hooke had already thought of the inverse ratio of the square of distances and started experiments on the surface of the earth for the purpose of finding out whether weight is the same at different altitudes. The differences were too small to be revealed by the experimental means at his disposal, and it was only in astronomical distances that favourable elements of comparison could be found. The clear conception of this fact is due to Newton, and this is the reason why his name is attached to the law of gravitation. The calculations, however, which he made in 1666 did not yield satisfactory results, and, without giving up his hypothesis, Newton did not think it expedient to publish his calculations.

It was in June, 1682, that he became acquainted with the results obtained by the French astronomer Picard with regard to the measurement of the terrestrial meridian. These new measurements modified considerably the data from which Newton had started. Taking up his calculations once more, he found to his joy that his conjectures agreed admirably with the results obtained by Picard. He was then able to affirm that weight and the motion of the moon can be explained by the same law, namely, the attraction exercised by matter upon matter, *in direct ratio of masses and in inverse ratio of the square of distances.*

Thus, in the discovery of universal gravitation, we find two types of mental process, one consisting in a new conception of a trivial fact, and the other being an accurate numerical verification made possible only by observations and accurate measurements (such as the arc of the meridian of Picard, and the observations and calculations previously made by Kepler). It may not be superfluous to point out that the accurate numerical verification was limited at first to a single example, that is to say it consisted simply in the agreement between *one* observed number and *one* calculated number. This agreement, however, was sufficient for Newton, who, in spite of his scientific caution, did not hesitate to formulate his law. This law thus became the starting-point of celestial mechanics, a complicated and often abstruse assemblage of analytical formulæ of interest to specialists only, whilst every cultivated man knows and admires Newton's law. It is, however, these formulæ of celestial mechanics which, by reason of their agreement with observation, have furnished us with an *a posteriori* demonstration of Newton's law, which rests upon thousands, nay, millions of accurate numerical agreements. This, however, was the work of several generations of astronomers, among whom we must single out two famous Frenchmen, Laplace and Le Verrier.

The Experiments of Cavendish.

However satisfactory the innumerable verifications of Newton's law by means of astronomical observations may be for mathematicians and astronomers, it is possible that direct experiments may be considered more convincing even though they are less accurate. If we leave aside for the present the famous experiment with the pendulum made by Richer at Cayenne in 1672, it was only in 1798 that Cavendish measured directly the attraction of a mass of matter by means of a torsion balance. He could at will increase or diminish the mass of matter the attraction of which he was endeavouring to measure, so that the effect to be measured depended upon the will of the experimentalist. This fact constitutes the principal difference between a scientifically conducted experiment and a simple observation. Cavendish's purpose, moreover, was not to demonstrate the law of gravitation, which no one any longer questioned, but to measure with precision what is termed the *constant* of gravitation, that is to say the constant factor whose product by the masses of two bodies and by the reciprocal of the square of their distances, gives us the exact measure of the mutual force they are exercising upon each other. The knowledge of this constant and of the intensity of gravity enables us to calculate the total mass of the earth and thus to weigh, by comparison, not only our own planet, but also the sun, the planets provided with satellites, and even certain double stars. The weight of the earth and of the sun is thus obtained by the observation of a weight variation of a few milligrammes, measured by means of an extremely fine pair of scales. Cavendish's experiments have been repeated several times during the nineteenth century, and—what would have been impossible for Hooke and for Newton's contemporaries — the difference existing between the weight of the same body at the foot and at the top of a high tower has also been directly observed. It suffices to employ for this purpose a balance with four scales, two at the end of each beam; on each side one of the scales is suspended beneath the other at a distance, for instance, of twenty metres. Now, if we obtain equilibrium by placing two bodies one to the right on the lower scale and one to the left on the upper scale, we shall find that the equilibrium will no longer exist if we change the position of the bodies, placing the one to the right on the upper scale and that to the left on the lower scale. This is due to the fact that the body on the right side has lost weight on account of its increased distance from the centre of the earth, whilst the body on the left side has increased in weight.

Gravitation an Isolated Phenomenon.

Thus, at the beginning of the twentieth century, universal gravitation presented itself as one of the best founded of scientific doctrines, established with almost absolute accuracy. There were, of course, some astronomical phenomena, the most important being the *secular motion of the perihelion of Mercury*, which did not absolutely agree with the predictions of theory, but the differences were extremely small. Besides, during the nineteenth century it had been noticed on several occasions that a more minute analysis and improved means of observation resulted in reducing similar differences to zero, so that the hope was still entertained that the law might be kept intact at the price of only an infinitesimal touching up.

There was, however, something rather strange in this phenomenon of gravitation, something that distinguished it from other physical phenomena. This was its utter immutability and its absolute independence of all external actions. Light is arrested by opaque bodies, deviated by prisms and lenses; electrical and magnetic actions are modified by the neighbourhood of certain bodies; gravitation alone remains always the same, and we have no means enabling us either to increase or to diminish it. Gravitation is indifferent to all physical circumstances and is not affected by the chemical nature of bodies. Radioactivity alone has furnished us with an analogous example of a property which is equally invariable, but this property, it must be remembered, is that possessed by a particular kind of matter, whilst the law of gravitation holds good for all matter. We must, however, call attention to the fact that, quite recently, the Italian savant Majorana has obtained certain results with regard to the absorption of gravitation by intervening bodies. These results, if confirmed, will be of far-reaching importance, but for the present the number of experiments is too small to be taken into consideration here.

Thus gravitation occupied among scientific laws a prominent but isolated place, having no connexion with the other laws. It had its particular domain where it reigned supreme, without fear of any possible interference, but it remained outside the constantly increasing close relationships established between the various parts of science. Many people are of opinion that these close relationships existing between the numerous scientific disciplines, even if they are not the whole of science, are at least its grandest expression. If the only worthy aim of human activity is the conquest of truth, we can hope to come near this inaccessible goal only by means of constantly widening syntheses. It is because it has effected the entry of universal gravitation into a more general conception of the world, particularly connecting it with electrical and luminous phenomena, that Einstein's theory of general relativity has been received with such admiration and passionate curiosity in scientific circles all over the world. We know that it was the Royal Society of London which, in 1918, took the initiative in organizing astronomical expeditions, the result of which was that the observation of the solar eclipse on

29th May, 1919, furnished us with a striking verification of the predictions which had been made on the strength of Einstein's theory. Newton's fame has nothing to fear from a progress which drags his law forth from its splendid seclusion and confers upon it a more lively rôle in the arena of scientific activity.

Centrifugal Force and Force of Inertia.

The theory of general relativity cannot be explained in detail with any accuracy otherwise than by means of a mathematical apparatus which is not within the scope of this book. But it may not be quite impossible to help to make the reader understand wherein the new conception of universal gravitation consists. The new conception has, indeed, consisted essentially in looking at familiar facts with new eyes; it is the history of Newton's apple over again. Already in 1672 the French astronomer Richer, during a scientific mission to Cayenne, had found that, in order to maintain correct timekeeping on the part of his clock, he had to shorten the pendulum by a line and three-quarters, that is by about four millimetres. The explanation of the phenomenon, which both Newton and Huyghens submitted to calculation and which raised mathematical difficulties solved for the first time by Clairaut, is as follows: the rotary motion of the earth round its own axis produces a centrifugal force in consequence of which gravity becomes less at the equator than it is at the poles. We are, of course, unable to measure this effect by means of a balance, as the action is exercised equally upon the weights placed upon both scales. If, however, we weigh a piece of metal with a dynamometer, that is to say by means of a spring which stretches more or less, we shall find that the same piece of metal weighs less at the equator than at the pole, and that the difference is about ten grammes for a weight of three kilogrammes. This is an example of centrifugal force, the nature of which we understand very well from simple experiments, such as may be made with a sling, for instance, but it is only a particular case of a more general law. If we wish to study the mechanics of a system which is involved in a general movement we are at liberty to set aside this general movement, on condition that we introduce supplementary forces, which we term forces of inertia. In other words, we may explain the experiments of Richer, or the diminution of the weight of a mass transported from the pole to the equator, in two ways. We may either admit that the earth turns, or on the other hand we may consider that it stands still, but that a vertical force directed upwards, the so-called centrifugal force, acts upon all bodies at all points of the equator (and also at all other points of the globe, but with an intensity varying according to latitude). Naturally, as Poincaré has pointed out, it is more *convenient* to suppose that the earth is turning, a convenience which, for Poincaré, is of the same nature as the convenience which, for the ordinary purposes of life, we find in admitting the existence of a perceptible world.

Gravitation a Force of Inertia.

Let us try, however, to look at the phenomenon of gravitation with the eyes of Einstein. We find that gravitation has every appearance of a force of inertia, inasmuch as all bodies are affected by it independently of their nature, and only according to their position in space and time. Now, we may look upon the centrifugal force manifesting itself at the equator of the earth as a property of space and time; for the rotation of the earth, which is a physical phenomenon, can be expressed mathematically by means of formulæ wherein space and time play a part. Equally well can gravitation be considered as an analytical property of a system of formulæ wherein space and time are combined in a new way. Mathematicians call this combination *a quadratic form of the differentials of four variables*, but we may call it more briefly, with Minkowski, " the Universe ".

Gravitation thus becomes a geometrical property of the universe, a geometrical property which in four-dimensional geometry is nothing but the generalization of what we term curvature in ordinary geometry. This notion of curvature is familiar to us; a billiard ball has a more pronounced curvature than an open umbrella, a plane has no curvature.

If we wish to indicate that a body is located at a certain point at a given moment, we can express it more briefly by saying that it is located at a certain point of the universe, and the manner in which it behaves depends solely upon the curvature (or rather curvatures) of the universe at this point.

Just as Newton's law has led us to very complicated mathematical calculations with regard to celestial mechanics, calculations which have indirectly furnished us with the most complete demonstration of the law itself, so the new ideas on gravitation lead us to an ensemble of calculations which will embrace all physical phenomena, the complete development of which will not be easy. The results already obtained, however, give us hope that confirmations will not be lacking.

With regard to demonstrations which are more direct and, as it were, immediately tangible, these are as difficult to find for Einstein as they were for Newton, but not more so. In both cases, the tremendous forces ruling the universe can be observed only in the form of extremely feeble residues. Luckily, however, the means of observation at our disposal are more perfect now than they were in the days of Newton, and we shall certainly not have to wait for more than a century for an experiment analogous to that made by Cavendish. Whether it be a question of the observation of the planets Mercury and Mars, or of the observation of the stars during a solar eclipse, it is these infinitesimal differences which have till now furnished us with the most convincing proofs. Here we have an admirable revenge for those conscientious scientists who, during centuries, have accumulated minute and accurate observations, have completed long and often irksome calculations, or have improved and perfected measuring instruments, enabling us to obtain an additional exact decimal for our knowledge of the universe. Certain superficial minds have looked upon these scientists as so many cranks, and considered it simply childish to take so much trouble in order to ascertain whether such and such a celestial phenomenon takes place at a few tenth parts of a second sooner or later. Is such a result, they asked, worth the trouble of thousands of nights of observations and volumes full of calculations? Once again, however, it is from this obscure decimal that the light bursts forth which enables us to understand the universe better, and even more than before to admire its perfect harmony.

CHAPTER I

Geometry and the Shape of the Earth

1. Origin of Geometry—Invariable Bodies.

Tradition traces the origin of geometry to the necessity imposed on the ancient Egyptians of discovering the former boundaries of their fields after the inundation of the Nile. The inundation having obliterated these boundaries, it was a problem of geometry to find them again. The very fact that the problem could be proposed was an implicit admission of the postulate that a solution was possible, or in other words, that fields could be found having the same area and shape after the inundations as before. It was assumed in fact that the ground possesses, approximately at least, the property of being inextensible and indeformable. Thus at the very beginning of geometry we already find that conception of a solid body which, as one may easily convince oneself, is inseparable from the very notion of space. In order to be able to locate objects in space, we must *know* these objects, that is to say they must, roughly at least, preserve their identity. If external objects (including our own body) were to appear to us as changeable as the clouds, whose shape is constantly being changed by the effects of wind and sun, it is probable that our conception of space would have been quite different from what it is, deprived as we would thus have been of any fixed landmark to which we could refer variable bodies. It would, however, be useless to try to imagine how space appears either to a bird that is moving in the midst of clouds or of a swarm of insects, unable to perceive a fixed point on earth or in the sky, or to a fish isolated in the midst of an ocean which is agitated by waves, and the surface of which is constantly changing its aspect. It is a question of ourselves and of our own conception of the universe, the result of the nature of our minds, fashioned as these have been by heredity and education. The value of science is in no way diminished by the observation that it is relative to man. What interest could there be for us in a knowledge quite unrelated to ourselves, if indeed the very conception of such a knowledge is not in itself contradictory?

2. Geometry of Position, and Metric Geometry.

Our acquaintance with the external world consists first of all in a fixed mental correspondence between certain spots and certain thoughts, a correspondence which seems to have been established even by the animals: the horse returns to its stable and the pigeon to its dovecot. When we try to analyse this knowledge of an immediately adjacent world, as possessed by an animal or a very young child, we can distinguish in it two essential elements, one corresponding to geometry of position, and the other to metrical geometry. Geometry of position embraces all those properties which may be described without our having recourse to measuring instruments, and which can be demonstrated by means of a roughly approximate drawing. When a passer-by, for instance, asks me the way and I tell him: " Follow this road and you will come to a thoroughfare where you will take to the right; you will then cross a bridge, after which you will find another thoroughfare where you will take to your left; you will then ascend a flight of stairs at the top of which you will have reached your destination "—all these indications refer to geometry of position. Our first knowledge of the external world is no doubt a knowledge of geometry of position, but even from the beginning a metrical element enters into it, that is to say a certain notion, more or less vague, of the relative value of distances. When I tell the passer-by who is asking me the way that he will reach a thoroughfare, I do not indicate the exact distance of this thoroughfare, but he would certainly look upon me as a mischievous wag if the thoroughfare of which I spoke were either less than a metre or more than a hundred kilometres from the place where we met. In the first case I would simply say: " Here is the thoroughfare ", whilst in the second I would warn him that the place he wished to go to was beyond the limits of a walk. Thus, even when we omit to give precise metrical indications, our language in itself presupposes some such vague indications. It is quite likely that the language which either animals or very young children make use of among themselves contains such indications when they think of the road they have to follow in order to reach a point familiar to them; it must include an estimate, at least implicit and approximate, of the distance. We shall have occasion to make analogous remarks when speaking of the shape of the earth and, later on, of the most general co-ordinates of the theory of relativity. It is therefore important to point out at once the nature of our most elementary representation of the external world; it is a representation of geometry of position, coupled with approximate metrical estimates.

3. Solid Bodies.

Geometry arose when man felt the need of greater exactness in these approximate metrical estimates. He found that there existed numerous bodies, solid bodies, whose lengths remained unchangeable, *in relation to each other*; it only required an easy and probably unconscious inference to conclude that this invariability was absolute, that is to say that the length of a wooden or metal ruler *always* remains the same. From this fact emanated the possibility of choosing once for all a unit of length and of estimating lengths by means of this unit. We shall discuss by and by the difficulties arising from the choice of such a unit of length, when it is a question of making measurements of very great accuracy. Such difficulties do not arise when we are content with only passable accuracy, such as has for a long time been sufficient for the requirements of man. Even to-day the common wooden metre, either rigid or flexible, suffices for the practical requirements of drapers, masons, and carpenters. On the other hand, the various parts of a motor-car engine necessitate much greater precision, and could not have been manufactured without the scientific progress which made such precision possible.

4. Cartesian Co-ordinates.

Once we have begun to define places by means of the objects occupying them, a slight effort of abstraction is sufficient to enable us to image a vacant place apart from the object which occupied it yesterday or will occupy it to-morrow. Take, for instance, an avenue planted with trees at regular intervals; I am able to number these trees and to define a certain spot in the avenue by saying that it is the spot where tree number 345 is to be found. Now supposing this tree to have been uprooted by a storm but the trees numbered 344 and 346 to have been left standing, it will be easy for me to find the exact spot occupied by tree number 345. With a little more care and attention I shall be able to do so even if only one tree in ten or even one tree in a hundred has been preserved. We thus easily arrive at the system of *co-ordinates*, the discovery and systematic introduction of which are due to Descartes. Let us imagine a vast plain over which a rectilinear road has been laid out; we shall call this road the axis of x, or the axis of *abscissæ*; all along the road milestones will enable us to tell the distance from a certain point fixed once for all, and called the *origin* of abscissæ. If we wish to go to a point A of the plain, we shall have to follow the road until we reach the point nearest to A, which means that from the given point A we let fall a perpendicular to the road and take the *foot* B of this perpendicular; if this point B of the road is at a distance of 3450 metres from the origin O, we shall say that the abscissa of the point in the plain is 3450 metres. If we subsequently have to walk 2250 metres perpendicularly to the road in order to reach the point A we shall say that the *ordinate* of A is 2250 metres; it may be either positive or negative, according as the point A lies to the right or to the left of the road. If, again, we sink a well 50 metres deep at the point A, we shall say that the bottom of the well has a *depth* equal to 50. The knowledge of these three numbers, the abscissa, the ordinate, and the depth, enables us to determine without any ambiguity the position of any point; these three numbers constitute the three Cartesian co-ordinates of this point. In order to define them, however, we have had to speak of rectilinear roads and of perpendiculars, and have thus had to assume a knowledge of certain geometrical notions. It would appear, therefore, that we are moving in a vicious circle if we take Cartesian co-ordinates as the ultimate basis of geometry. We shall soon see, however, that it is perfectly legitimate to give co-ordinates this position, but a few preliminary explanations are needed.

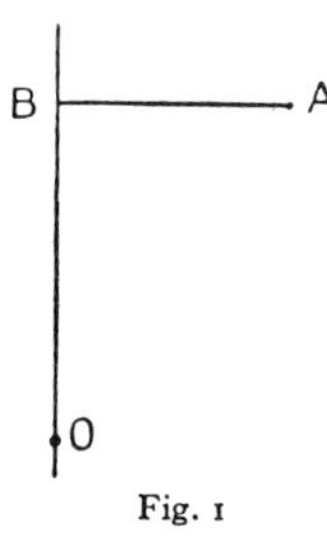

Fig. 1

5. The Postulates of Euclid.

Let us retrace our steps and go back twenty centuries, from Descartes to Euclid, and try to understand Euclid's conception of space. The essential element in Euclid's system is the straight line, to which certain properties are attributed *a priori*, properties considered as axioms. In addition to these there is also one property placed apart, the famous Euclidean postulate or axiom of parallels, according to which through an external point we can draw only one parallel to a straight line. The fact that he was led to put this property in a place apart from the others is, perhaps, the most brilliant proof of Euclid's inductive genius. We have no intention of entering here into a detailed discussion of his axioms and postulates; we admit them *en bloc*. When this is done, it is easy to present them in various forms which appear equally intuitive. One of the simplest of these equivalent sets of axioms and postulates amounts to the assumption of the possibility of dividing a plane into equal squares, and space into equal cubes. If we confine ourselves to the plane, the fact that we are able to pave a plane surface with square paving stones, one equal to the other, without leaving an empty space, is one of the simplest and earliest observed of the facts known to artisans. When we say that the paving stones are square it means that their sides and their angles are all equal among themselves, so that each paving stone can be placed in four different ways (and even eight if we turn it upside down) in the cell reserved for it among the other paving stones, supposed to be in their allotted places.

Another alternative, equally intuitive, to the Euclidean axioms and postulates, is the statement of the possibility of making a drawing on a reduced scale which shall have exactly the same shape as a given drawing or sketch. This statement had already been implicitly admitted by the cave man when he traced on the sides of the rocks those drawings representing contemporary animals which have been discovered in prehistoric grottos.

If we admit as an experimental fact the one or the other of the above statements, namely the possibility of a regular arrangement in squares, and the possibility of a drawing reproducing the contour or the shape of objects, we at once admit Euclidean geometry, *at least for objects on our own scale*. We can then accept and make use of the definitions of Cartesian co-ordinates. With regard to geometry in space, we shall have to admit the possibility of dividing space into equal cubes, or the possibility of constructing models reproducing on a reduced scale the exact shape of objects.

6. **Analytical Geometry and M. Jourdain's Prose.**

We can go a step further and consider Descartes' co-ordinates as practically *equivalent* to Euclid's postulates. This manner of proceeding supersedes for our purpose very conveniently the long discussions with regard to the axioms of geometry in which numerous philosophers and geometricians indulged at the end of the nineteenth century. I do not in the least lack appreciation of the interest offered, from certain points of view, by those studies in which one examines a series of questions as to what would become of abstract geometry if we were to leave out such or such an axiom. But from the realistic point of view, which is the one we adopt, if it is simply a question of formulating the axioms of Euclidean geometry, the simplest thing to say is that this geometry is, after the explanations of Descartes, exactly equivalent to a study of the qualities of three numbers which are the three rectangular co-ordinates. This is not the place to develop the elements of analytical geometry from this point of view. What we are anxious to do is to define, as exactly as possible, the different points in space, and for this it is sufficient to specify three rectangular axes and a unit of length.

In consequence of an inexplicable pedagogical routine there are many civilized men who, in spite of a primary and even higher education, are still ignorant of, or believe that they ignorant of, rectangular co-ordinates. As a matter of fact, they are as little ignorant of rectangular co-ordinates as was M. Jourdain of prose. Unconsciously he had been in the habit of expressing himself in prose from his very childhood, and marvelled greatly when he learned the name of this ordinary language. To our contemporaries who imagine they know nothing of this sort of mathematics, but are cultivated enough to read the daily papers, analytical geometry and Cartesian co-ordinates are what prose was to M. Jourdain. They may be alarmed at first by these mysterious terms, but before long will be greatly astonished to learn that when they are looking at the graph representing the rise or fall of the price of bread they are making use of analytical geometry and of Cartesian co-ordinates. They are doing the same thing when they are looking for a street on the plan of a town divided into small squares by means of numbered strips. The list of the names of the streets gives them the co-ordinates, that is to say the numbers of the strips whose intersection indicates the square in which the street is to be found.

7. Analytical Geometry—Space on the Human Scale.

The use of Cartesian co-ordinates on a scale near the human scale offers no practical difficulty; whether it be a question of a plan of a town, or of a cadastral map, or of a map of the General Staff, we always have two rectangular axes, as, for instance, the edges of a sheet of paper upon which the map is drawn; these axes correspond to straight lines fixed on the ground, and distances on the map stand in a fixed proportion, which is the scale, to distances on the ground. The third axis, at right angles to the two first, is simply the vertical line. In the representation of any part of the earth's surface in a plane map we mark at the side of each notable point a number which is equal to this third co-ordinate; it represents the height of the point above a fixed horizontal plane, taken as the origin of heights. If we are in the neighbourhood of the sea, this fixed horizontal plane will generally coincide with the mean sea-level. Thanks to these arrangements and to the chosen data, to which we must also add an unchangeable standard of length (which for us is the metre), we are able to fix, in a precise manner and by means of three numbers, the position of a point which is not too far away from the origin of the system of axes chosen. We can proceed in this way: the origin will be materialized on the earth's surface; it will be, for instance, the top of a steeple, or, more precisely, the point where the vertical line passing through the top of the steeple pierces the horizontal plane at sea-level of the place; the axis of z will be this vertical line, the axis of x will be the horizontal line directed towards the east, whilst the axis of y will be the horizontal line directed towards the north. The point in space for which $x = 345{\cdot}4$ metres, $y = 248{\cdot}3$ metres, $z = 42$ metres is then completely determined without any ambiguity. We may therefore say that we have a perfect knowledge of the space in the neighbourhood of the steeple, since this knowledge is reduced to our acquaintance with three numbers, and since the methods of algebra and analytical geometry enable us to study with ease the properties of this system of numbers.

8. Number knows no Limitations.

Whilst, however, the introduction of numbers for the purpose of measuring and representing space is a great simplification, it is also a source of difficulties. Number knows no limitations, either from the side of the infinitely great or from the side of the infinitely small, and the facility it offers for generalization is too great for us not to be tempted by it. The Greeks, too, have shown us the way; but let us keep to our rectangular axes. The metre being our unit, we are able to find, without any difficulty, a point whose co-ordinates are 345·4, 248·3, and 42 respectively. From a theoretical point of view there should be no greater difficulty if we take as co-ordinates 100,000,000, 250,000,000, and 300,000,000, or 0·000001, 0·0000002, and

0·00000004. Do we, however, *possess* these new points, that is to say, do we have a clear conception of them? Are we able to construct and reach them, or, at least, imagine the methods by which such points can be constructed and reached? These questions deserve the most careful consideration. We shall first study them so far as they concern lengths considerably exceeding human dimensions, reserving consideration of the difficulties connected with very small lengths till Chapter IV.

9. Preservation of Landmarks.

We have said that if we wish to define a system of rectangular co-ordinates to which we refer a small terrestrial space, it is sufficient to have given a fixed point, say the top of a steeple, and a fixed horizontal plane, which will be the mean plane of the sea-level if we are in the neighbourhood of the sea. These data, and the geographical directions of north and east, completely determine the axes. But what will happen should the steeple be destroyed? Will it be possible to discover the position it occupied? Here we have a real difficulty which sometimes actually presents itself when we try to interpret geographical information left by the ancients. If a town has disappeared without leaving any trace, how can we discover the position it once occupied on the surface of the earth? A solution of this question is offered by the use of geographical co-ordinates, and although this solution is only approximate and incomplete, it should be studied carefully on account of the information it furnishes.

10. Geographical Co-ordinates.

We know that we are able to specify the position of a point on the surface of the earth by means of two co-ordinates, latitude and longitude. Supposing the earth to be strictly spherical in shape, we may define latitude as the angular value of the arc of the meridian separating the given point from the equator, and longitude as the angle formed by the meridian of the place and a fixed meridian taken as origin (the prime meridian). In these definitions we notice the approximate and incomplete nature of geographical co-ordinates. In order to be able to define the position of a point with any exactness, we must be able to make angular measurements with a precision so great as to be practically unattainable. A sexagesimal minute of arc on the arc of the meridian corresponds to more than 1800 metres, so that the tenth part of a second of arc will correspond to a little over 3 metres. If we wish distances to be correct within 3 millimetres, a length easily appreciable by the naked eye without the employment of special instruments, we shall have to estimate angles to the ten-thousandth part of a second of arc, which it is quite impossible to do with the instruments by means of which we measure angles (theodolites). So much for latitude; as for longitude, pretty much the same can be said as to its approximate character, but the definition of longitude is also incomplete in another way, inasmuch as it requires the fixing, once for all, of a prime meridian. How can we define this in such a way as to enable us to find it again after the lapse of many centuries? We might try to obviate the difficulty by defining several natural landmarks whose mutual distances are well known and whose respective longitudes are equally well known. We might then hope that at least one of these would be extant; and should several be extant, that is to say have undergone only very slight modifications, then we could reduce errors by taking an average.

The provisional geometrical definition of latitude which we have given is, however, doubly inexact. On the one hand, the earth is not spherical in shape, whilst, on the other, the action of centrifugal force slightly modifies the direction of gravity, and there is an advantage in defining latitude in terms of the direction thus modified. The correct definitions of latitude and longitude, definitions which make no assumption with regard to the shape of the earth, will therefore be as follows. Latitude is the complement of the angle which the vertical of a place makes with the axis of the earth (the line of the poles); longitude is defined in *sidereal time* as the interval of time separating the crossing (transit) by a certain star of the meridian of the place and its crossing of the prime meridian.[1] More precisely, the interval of time elapsing between two consecutive transits of one star across the meridian of the same place, will be defined as equal to twenty-four sidereal hours, and will correspond to 360°; if the interval elapsing between the transit of the same star across the meridian of two different places is the twenty-fourth part of this, we shall say that the interval is a sidereal hour, and that it corresponds to a distance of 15° in longitude. We see introduced here for the first time the notion of time in the definition of a position in space. We will not speak here of the difficulties raised by the introduction of such a notion, and assume that we possess well regulated clocks and chronometers, and that the use of wireless will enable us to secure perfect agreement between two chronometers at different spots on the earth's surface. A little later we shall discuss the objections and difficulties raised when we examine these simple notions more closely, but for the moment we will consider them as sufficient for our present purpose.

11. Geodetic Measurements.

Independently of or concurrently with the use of geographical co-ordinates, one may propose to connect the various landmarks on the earth's surface by means of direct measurements. The methods employed for the purpose of carrying out such measurements constitute *geodesy*. It

[1] Of course, in the case of a circumpolar star, only *superior* transits (for example) will be considered.

was not until the seventeenth century that precise geodetic measurements were first undertaken. The method adopted, a method which has changed but little, consists in measuring first of all a rectilinear base of a few kilometres, the measurement being done by means of rods placed end to end. Recently it has been found advantageous to substitute metallic wires for rods, but this is not the place to insist upon either details or technicalities. We may mention, however, the advantages offered by the use of the metal *invar*, the expansion of which is practically nil (Ch.-Ed. Guillaume). The base having been measured, we consider it as one of the sides of a triangle, and we cover the whole region to be measured with a network of triangles having two by two a common side. The measurement of the angles of these triangles enables us step by step to determine all the elements, starting with the side directly measured. For the purpose of verification we measure directly a new base, as far as possible from the first, that is to say a side already calculated of a triangle far from the original triangle; and agreement between this measurement and the previous calculations will prove the accuracy of the intermediate operations, such agreement being much more likely to be due to the exactness of the work than to mere chance. Triangulation thus consists in calculating all the elements of a polyhedron with triangular faces the vertices of which are well determined points on the earth's surface (such as steeples, signposts on mountain summits, &c.). The study of the shape of the earth in the light of the results obtained by means of triangulation constitutes the principal problem of geodesy. The immense extent of the oceans and of the unexplored regions of the earth in comparison with the continents accessible to us, make the geodetic problem as a whole a very difficult one to solve. As a matter of fact, when it is a question of determining the difference of longitude between Paris and Washington it is to astronomical methods, supplemented by chronometry and wireless, that we must have recourse.

On the other hand, we must point out that the method of measuring a *rectilinear* base, as well as the process of triangulation by means of measurements of the angles of large triangles defined by light signals, presupposes that we have adopted as our definition of a straight line that it is the path of a ray of light. Moreover, the measurement of the base presupposes an exact definition of the unit of measure and a very minute study of expansions, which the use of the metal invar may simplify, but does not completely supersede. It is therefore the entire science of physics that we shall have to study if we wish to get a clear idea of the means to be employed when we try to make any but the roughest of space measurements. Thus we need not be astonished at the mediocrity of the older geographical maps. The errors in these maps are so great that they strike even the ordinary reader who has no other knowledge of modern maps than that vaguely derived from the fact of his having consulted them from time to time without making any special study of the subject. It might be interesting and instructive to give a brief sketch of the history of the progress achieved by both geodesy and cartography, but such a history would be beyond the scope of this volume. It will therefore be sufficient to call attention to a few essential points.

12. The Unit of Length and Richer's Pendulum.

Let us first say a few words with regard to the choice of the unit of length. The oldest idea, which is also the simplest one, consists in taking as a unit the length of some carefully preserved standard, copies of which are made for use in ordinary practice, so as not to wear out the standard itself. As a matter of fact, the standards used by the ancients have not been preserved, and the reconstruction of their measure of length is consequently only approximate. That it has been possible at all we owe to historical documents containing information with regard to the dimensions of some edifice the ruins of which are still extant, or with regard to the distance between two towns the ancient sites of which are well known.

Thus, when in the seventeenth century geodesists undertook to make accurate measurements of certain arcs of the meridian, they sought to give the results of these measurements in a form which should not be dependent on the preservation of the *toise*, the legal measure in France at that time. For this purpose Cassini fixed the relation in which a pendulum beating seconds stands to this standard measure. This means, in short, that he took as standard of length the length of his pendulum, that is to say that he specified the unit of length by means of the unit of time. As we have said above, we shall reserve the discussion of the difficulties involved in a precise definition of a unit of time. There is, however, in the above definition of the unit of length a foreign element of quite a different nature from length; it is the value of the intensity of gravity at a given point, a physical quantity on which the length of the seconds pendulum depends.

We usually designate by the letter g the number measuring the intensity of gravity, that is to say the acceleration with which heavy bodies fall in space; in other words, it is the speed acquired after one second by a heavy mass leaving its point of rest and falling freely in space.[1] The value of g is far from being constant over the earth's surface. It is smaller at the equator than it is at the poles, on account of centrifugal force and the shape of the earth. This important fact was noticed for the first time by the astronomer Richer in 1672, during his scientific journey to Cayenne. He had to shorten the pendulum of his clock by a line and three-quarters (about 4 millimetres) in order to

[1] This definition implies, of course, that the unit of time has been defined; we choose the second, i.e. the 86,400th part of the mean solar day.

make it continue to beat seconds. Richer's observation was interpreted by Newton and Huyghens as a consequence of the shape and the rotation of the earth. Precise theoretical results, however, were deduced only slowly and by successive steps. Newton assumed that the earth was homogeneous—we know now that the density increases from the surface to the centre, the density of the upper layers being about half of the mean density. He also assumed, without demonstrating the statement, that the figure of equilibrium of a homogeneous fluid mass subject to the law of attraction and turning round an axis is a flattened ellipsoid of revolution. This theorem of mechanics was demonstrated in 1737 by Clairaut, and in 1740 by Maclaurin. In 1743 Clairaut, in his *Théorie de la figure de la terre*, solved the same problem for the case of a spheroid constituted of homogeneous layers of variable densities. He demonstrated in this general case the formula already assumed by Newton:

$$g_L = g_0(1 + \beta \sin^2 L),$$

g_0 being the value of g at the equator (latitude 0) and g_L its value at a point of latitude L. Clairaut showed also how the knowledge of the coefficient β, added to the knowledge of centrifugal force at the equator, enabled us to calculate the flattening α of the earth, that is to say the ratio to the greater axis of the difference between the greater and the smaller axes, or $\alpha = (a - b)/a$.

The correct value of the flattening was found in this way to be nearly 1/300. Newton had found 1/230, and Huyghens, by reasoning resting upon wrong premises, 1/578. The equatorial radius of the earth (half the greater axis) being in round numbers 6300 kilometres, the polar radius (half the smaller axis) is shorter by about 21 kilometres. The difference therefore between the earth and a sphere is not quite unimportant. We must not forget that the experimental origin, if one may call it so, of these 21 kilometres is to be found historically in the difference of a line and three-quarters, or 4 millimetres, noticed by Richer (the line being equal to 2·256 millimetres).

13. The Metric System and International Standards.

Let us, however, return to the unit of length. The progress achieved in the course of the eighteenth century in our knowledge of the shape of the earth, thanks especially to the expeditions organized by the Academy of Sciences in Paris to the polar and equatorial regions (Maupertuis to Lapland, Godin, Bouguer, and La Condamine to the territory, then part of Peru, but now occupied by the Republic of Ecuador),[1] suggested the idea of taking the terrestrial globe itself as the standard of length. The metre was then defined as the forty-millionth part of the length of the earth's meridian. Thus the initiative in substituting for national measures a standard whose definition is general enough to suit all nations was taken by the French. The metric system, under the name of the C.G.S. system, has actually been adopted by scientists and electricians of all nations, whilst for commercial purposes it is now used by nearly all civilized nations. Very likely Great Britain too will end by adopting this system, perhaps a few centuries after China and Japan.

The improvements made in geodetic measurements should naturally have caused us to modify slightly the definition of the metre with every new advance realized in our knowledge of the earth. But it was found preferable to define an unchangeable standard metre differing but slightly from the theoretical metre of the founders of the metric system. Thus the definition of the standard of length was settled by means of an arbitrary bar deposited in a fixed place, namely at the head-quarters of the International Bureau

[1] The measurement of an arc of the equator has again been undertaken at the beginning of the twentieth century by a mission consisting of French officers organized by the Geographical Army Service.

of Weights and Measures (at the Pavillon de Breteuil at Sèvres). Copies as exact as possible of the international standard were made for every country which had adhered to the convention on the metre, and considering the number of these copies scattered all over the face of the earth, it may not be too rash to believe that the standard metre with the precision it actually possesses will last as long as our civilization. Now how great is this precision? It does not attain the thousandth part of a millimetre,[1] that is to say the millionth part of the fixed length. We might, perhaps, increase it slightly by assuming as a convention that the metre is equal not to one fixed standard but to the average of various standards copied one from the other. Theoretically these copies are equal among themselves, but considering the physical qualities of the metals out of which the standards were manufactured, it does not seem possible to admit that these standards could be defined with a precision exceeding one ten-millionth part, that is to say assuring an accuracy beyond the seventh decimal. We find this seventh decimal an almost insurmountable obstacle in many other physical measurements.[2]

14. The Metre in Terms of Wave-lengths.

Physicists, however, were far from satisfied with a standard of length the definition of which depended on material objects preserved for the purpose, such as the official metres made of iridized platinum. As soon, therefore, as the progress of experimental optics made it possible, thanks to the splendid researches of Michelson on inter-

[1] To understand the degree of precision attained, the reader may consult with great advantage the very interesting note in which M. Ch.-Ed. Guillaume, Director of the International Bureau, discusses from a comparative point of view the modifications in the various standards at the end of a period of thirty years. It will be seen that difficulties present themselves of which we have not been able to speak here (*Comptes rendus de l'Académie des Sciences*, 27th Dec., 1921).

[2] It would be interesting to discuss from this point of view the experiments of high precision by which M. Benoit has determined the weight of a cubic decimetre of water.

ference, the metre was evaluated in wave-lengths of a definite line in the spectrum. We shall return a little later to the theoretical difficulties such an evaluation involves; suffice it for the moment to state that its precision attains the precision of the definition of the international metre itself. It would therefore be quite legitimate to say at once that the wave-length of a definite line of the spectrum is chosen as the standard of length. Such a definition may, perhaps, entitle us to hope for one or two additional decimals in the measurement of lengths, as the methods of physical optics are gradually perfected. This hope, however, will only be realized at the cost of very great efforts, and the result obtained, however great its practical and theoretical importance, will in no way invalidate the consequences one may draw from the fact that our actual knowledge of the earth is limited by the seventh decimal.

15. The Figure of the Earth.

Supposing the standard of length defined, then in order to be able to specify the different points of space, that is to say to know space practically, we have to choose a system of axes to which these points can be referred. We have already indicated how a local system of axes can be chosen. To connect these diverse local systems of axes among themselves is neither more nor less than the general problem of geodesy: to determine the shape of the earth. We have already referred to the progress of geodesy during the eighteenth century, when discussing the definition of a standard of length. But what is even more important than the history of the various geodetic expeditions is the principle of the method which enables us to determine the earth's exact shape and dimensions. The only satisfactory geometrical operation is that suggested by Marcel Brillouin in 1900;[1] this, however, unfortunately seems to be too difficult to carry to a practical conclusion. We

[1] *Revue générale des sciences.*

would have to choose points upon the earth's surface (such as geodetic stations, mountain tops, isolated islands in the ocean, &c.) numerous enough to make it possible to connect them by a direct triangulation. The measurements of the angles formed by the straight lines directly joining these points two and two (or at least those pairs of the points which would be mutually visible) would enable us to calculate all the elements of a polyhedral surface, exactly resembling the polyhedron the tops of which are formed by the chosen points. It would then be possible to try and find out whether there existed a simple geometrical surface, such as an ellipsoid, sufficiently near to the polyhedral surface to be considered, to a first approximation, as representative of the shape of the earth. We could then, to obtain a second approximation, study the distances separating the actual points of the earth's surface from the nearest points of the ellipsoid. It might be objected to this method that it takes no account of altitude, a physical element the importance of which cannot be denied. It would, however, be easy to obviate this objection by connecting up with the original polyhedron—the vertices of which are, of course, the highest points of the earth's crust—a secondary polyhedron having its vertices near the sea-level. Be that as it may, the moment has certainly not yet arrived when we can thus determine the shape of the earth by purely geometrical operations. The chief results actually achieved are these: we have made some measurements of arcs of the meridian, and also some measurements of arcs of parallels; the angular amplitudes of these arcs are determined by means of astronomical observations (supplemented, in the case of arcs of meridian, by a knowledge of the position of the vertical, and, in the case of arcs of parallels, by a comparison of chronometers); these arcs fit more or less accurately the various ellipsoids suggested by the geodesists.

16. The Earth regarded as a Level Surface.

There is some advantage in substituting for these geometrical methods of determining the shape of the earth another method, which is not purely geometrical, but depends partly upon mechanical principles. It is the problem of determining the mean surface of the oceans when this is supposed to be continued under the continents. One might ask, indeed, whether the problem in this form is properly stated, seeing that suppression of the elevated plateaux and mountain chains, which play a part in forming the figure of the earth, is not within our discretion. We would therefore have reason to ask with M. Brillouin whether we suppose the surface of the oceans to be continued under the continents by tunnels or by a system of open trenches. In the latter case the suppression of rocky masses necessary for the excavation of the trenches would slightly modify the level obtained when tunnels are used. It seems to be agreed, however, that the correct definition of the "geoid" is the surface of the oceans fictitiously prolonged by means of tunnels underneath the continents. This surface is what is called in mechanics a *level surface*, an expression which seems natural enough considering that its definition turns on the idea of a common level of a unique ocean supposed to cover the entire surface of the globe. The verticals are perpendicular to the level surfaces. For theoretical reasons, however, many geodesists, instead of the actual surface of the seas, have thought it preferable to deal with a level surface which would be situated twenty kilometres below or ten kilometres above. For the calculation of these new surfaces they call in, as far as possible, the aid of geological knowledge of the earth's crust, thus obviating the difficulties arising from the irregularities of this crust. We must, however, admit that this method allows a great deal of arbitrariness, and we shall confine ourselves to considering the surface of the oceans. Had this surface been in equilibrium, it would have been quite correct to consider the ensemble of the seas, supposed in mutual communication, as constituting one unique level surface. In reality, however, equilibrium is not established, since there are permanent currents in the oceans. We do not yet know very exactly what are the differences of level between the different oceans or even between distant parts of the same ocean, and it would not be easy even to give a definition of what we mean by these differences. We know, however, that the differences are small, so that the error made in considering the ensemble of the oceans as one unique level surface is less than some other more important errors. The surface thus specified is the *geoid*, and the result, which appears to be established although disputed by certain authors, is that the geoid differs but little from a certain flattened ellipsoid of revolution. We now fix upon one definite ellipsoid of revolution, differing but little from the true *geoid*, and select this as the approximate geoid to which we shall refer the various points of the earth's surface. Once the terrestrial ellipsoid has been specified, instead of taking particular axes having their origin at a point on the earth's surface, and directions depending on the position of this point, it will be natural to choose as origin of co-ordinates the centre of the ellipsoid, and as axis of z the axis of revolution of this ellipsoid. According as we do or do not want these chosen axes to follow the earth in its movement of rotation about its own axis, we specify the axes of x and of y in the one case by means of a point on the earth's surface (we may say, for instance, that the plane of x and z passes through Paris Observatory), or, in the other case, by means of astronomical directions. We shall return further on to the choice of these astronomical directions.

17. Variation of the Poles—Tides in the Earth's Crust.

We have thus reached, not without some trouble, the specification of a system of axes connected with the earth, defined in a symmetrical way and as unchangeable as the earth itself. As a matter of fact, there are still many difficulties, the chief of which we can only mention without going into details.

First of all the position of the poles varies. To be more precise, the points on the earth's surface through which passes the imaginary line round which the earth is rotating are not rigorously fixed, their displacement amounting in the course of a year to several metres. These displacements have been investigated, thanks to an astronomical study of the variations of latitude. It is clear that if the North Pole draws nearer to Paris on the Paris meridian, the latitude of Paris will increase, whilst the latitude of a point in the Behring Straits, situated at a distance of 180° longitude from Paris, will diminish to the same extent. A displacement of the pole to the extent of about 1852 metres corresponds to a variation of latitude of one minute of arc, whilst a displacement of about 31 metres would correspond to a variation of one second of arc; and as a matter of fact variations of latitude have been observed amounting to a few tenths of a second. The pole describes a somewhat irregular curve winding round an approximately fixed centre in successive spirals; in about fifteen months a turn is completed, but the phenomenon does not repeat itself strictly periodically. It is therefore impossible for the moment to define with precision a point which one could call the mean position of the pole, a point through which one might be tempted to let pass one of the system of fixed axes connected with the earth.

We could also speak of the more regular and better known variation of the earth's axis of rotation in relation to the fixed stars. This is what we call the phenomenon of the *precession of the equinoxes*, the consequence of which is an apparent displacement of an imaginary point on the starry sphere, a point at present in the neighbourhood of the pole star, round which the stars seem to be rotating. This imaginary point describes in about 26,000 years a circle among the fixed stars, so that in a few thousand years the star now called the pole star will no longer deserve that name. In 13,000 years its distance from the pole will be more than 45°, and it is the star Vega in the constellation Lyra that will then have the right to the name of pole star.

But let us come down to earth again. An important problem bearing on the determination of the position of points on the earth's surface relative to axes having their origin at the centre of our planet is that of the tides of the earth's crust. It was Lord Kelvin who, in 1863, formulated the hypothesis that the rigidity of the earth ought to be intermediate between that of glass and that of steel; the solid portion of the earth must consequently be deformed under the influence of the attraction of the sun and the moon, just as the oceans, in consequence of the same influence, produce the phenomenon of the tides. It is clear that if the solid portion of the earth's crust could be as easily put out of shape as water (if it possessed, for instance, an elasticity analogous to that of india-rubber) we would not observe at the shore of an ocean the phenomenon of the tides, since the shore itself would participate in them. (We must, however, make a reservation with regard to exceptional tides in straits, such as the English Channel.) We might therefore try to discuss the tides of the earth's crust by comparing the tides actually observed at ocean shores with the theoretical tides which would be produced if the terrestrial globe were perfectly rigid. This calculation of theoretical tides is unfortunately a very difficult one, on account of the complicated configuration of the shores; it is only in very particular cases that it has been possible to attack the question by calculation.[1]

The English astronomer G. H. Darwin, however, found it possible to obtain some results by applying his calculations to the monthly and bi-monthly tidal waves, which are less important than the daily or half-daily waves, but upon which the influence exercised by the shores is much smaller. It was, however, only at the beginning of the twentieth century that Hecker at Potsdam succeeded in determining with some precision the tides of the earth's crust, by means of observations of pendulum oscillations, using the instrument called the horizontal pendulum. The experiments of Michelson at Chicago have given results agreeing on the whole with those of Hecker, but differing in an important point, although the difference may be due to local circumstances. Whilst at Potsdam the rigidity of the earth appeared to be different according as it was observed in the direction north-south or in the direction east-west, no similar difference appeared at Chicago.

The general result derived from the study of the tides of the earth's crust is that they are about a third of what they would be if the earth possessed no rigidity whatever and yielded completely to the attraction of moon and sun. In the latter case the vertical would not deviate under the influence of the field of gravitation produced by the moon and the sun, whilst, on the contrary, its deviation would be much more pronounced if the earth's rigidity were absolute. As a matter of fact, the deviation observed is about two-thirds of the theoretical variation in the case of a rigid earth.[2]

[1] See the thesis for the doctorate of a young astronomer and mathematician, Blondel, whose work showed the most brilliant promise, but who unhappily was one of the victims of the war of 1914. In his thesis Blondel applied to the tides of the Red Sea the method indicated by Poincaré in his *Leçons de mécanique céleste*.

[2] In Chapter VII we shall explain why, if the solid earth yielded to gravitation, the pendulum would not disclose the existence of this gravitation.

To complete our account, we should have to explain the connexion existing between the two phenomena we have just described, namely the displacements of the pole and the tides of the earth's crust, but we must be content with a mere mention of the relationship.

To sum up, we see that all attempts made for the purpose of finding fixed axes upon the earth have succeeded only in an approximate way. The problem presents two sorts of difficulties, human difficulties and natural difficulties, the first resulting from the imperfection of our instruments and the insufficiency of the methods employed, whilst the latter are due to the instability of the earth's surface and the consequent impossibility of seizing with certainty on any fixed point in a world where everything is in a state of change. It seems reasonable, however, to hope for some progress in our struggle with the two difficulties mentioned. On the one hand, our methods of investigation are gradually becoming more perfect, and the resources at the disposal of science are increasing in proportion as the importance of scientific researches is being more appreciated; whilst, on the other hand, the study of the various types of variation of the earth's surface is making constant progress. We may thus hope to be able, with the help of averages, to define with ever-increasing precision elements forming a definite fixed system. Whilst, however, we may expect some progress, we must reconcile ourselves to the fact that this progress will always remain limited. Instead of the seventh decimal we may, after some centuries of effort, succeed in fixing with precision the eighth or even the ninth decimal. This is very little compared with absolute accuracy, but it may mean a great deal from the point of view of our knowledge of nature; we have only to think of Richer's line and three-quarters, which enabled us to infer the flattening of the earth.

18. The Scientific Value of Exact Measurements.

There is no reason, therefore, for despising the patient and sometimes obscure labours by means of which geodesists are steadily making our knowledge of the earth's shape and dimensions a little more exact—a little less false, some ill-natured people may say. Nor should one forget that it was for the purpose of dealing with certain problems of practical geodesy that the mathematical Calculus of Errors was originally devised by Laplace and Gauss. Here, too, it is easy to criticize, especially since the use made of excellent principles has frequently laid itself open to well-founded criticism; but must one consider the theory of logarithms as inexact because some careless person has made a mistake when using a table of logarithms? We must not expect the theory of errors to take the place of well-constructed measuring instruments or of the conscientiousness of the experimentalist. Still less must we expect this theory to make up for lack of precision in the theoretical considerations which have suggested this or that measurement. But once we know whither we are going, and good instruments in the hands of conscientious observers being granted, there is no doubt that a methodical criticism of results by the classical methods of the Calculus of Probabilities will enable us to make the very best of these results, and furnish us with one exact decimal more than the simple calculation of an average or a rough graphical procedure could have given us. And, we cannot repeat it too often, one exact decimal more may be the beginning of a great scientific discovery.

We have seen that, for the purpose of defining terrestrial axes—that is to say, of securing command of the space which is, so to speak, within our grasp—purely geometrical methods are not sufficient, even when supplemented by the hypotheses concerning the propagation of light which are implied in the laying out of a rectilinear base and the measurement of the angles of geodesic triangles. We have had to call to our aid mechanical ideas, and make use of the pendulum and the measurement of time; astronomical observations, too, necessitating the use of chronometers regulated by means of wireless, were found indispensable. We have thus been able to express the vague notion of space in numbers only by depending upon results which are foreign to the science of space. To these results we shall have to return again. The same circumstances will arise in an even more striking way, in the next chapter, where we shall endeavour to define those portions of space which are inaccessible to us and are known only through the stars which occupy them.

CHAPTER II

Space and Time in Astronomy

19. Modern Astronomy is not Geocentric.

We cannot attempt to retrace here the history of astronomy. Leaving aside the gropings of the ancients, however interesting they may be for a history of human thought,[1] let us come at once to modern times, to Copernicus, Kepler, Galilei, and Newton. The earth is no longer the centre of the universe, but a planet describing in the course of a year an elliptical orbit round the sun, whilst rotating at the same time round its own axis in twenty-four hours. The planets, too, describe known orbits, whilst the distances of the stars are immense in comparison with the dimensions of the solar system. Light, which comes from the sun in eight minutes,[2] takes several years to come from the nearest star; and as for certain nebulæ, the time which has elapsed since a ray of light reaching us left them is counted by millions of years. If we limit ourselves first to the solar system, how can we succeed in defining with precision the points in space outside the earth?

20. The Distances of the Planets are deduced from Newton's Laws.

One evident remark forces itself upon us at the outset. The only precise observations which astronomers can make

[1] See the excellent book by M. Jules Sageret: *Le système du monde* (from the Chaldæans to Newton). (Alcan.)

[2] The calculations of Kepler and Newton assumed the propagation of light to be instantaneous.

are observations with regard to angular directions. We project the planets, the sun and the moon, upon the practically unchangeable field constituted by the fixed stars, that is to say upon a fictitious sphere where the fixed stars serve as landmarks; but to fix their distances from the observer (and, consequently, from the centre of the earth) is much more difficult. It is true that observations of apparent diameters give us some vague indication with regard to comparative distances, but such measurements are entirely lacking in precision. The most considerable apparent diameters, those of sun and moon, hardly exceed 30 minutes, or 1800 seconds. A second of arc thus represents only a precision of 1/1000, that is to say, enables us to obtain the third decimal; and the observation of a tenth of a second would give us, at the utmost, the fourth decimal. The apparent diameters of the most important planets, Jupiter, Venus, Mars, are much less than those of the sun and the moon,[1] so that an observation of the tenth part of a second would give us only the third decimal. Thus, if we were reduced to purely geometrical measurements, our information with regard to the position of the stars nearest to us would be far from exact. These measurements, however, were sufficient to allow Kepler to state as probable his celebrated laws, afterwards confirmed in a most convincing manner by Newton's analysis. With the help of observations of the periods of revolution of the planets round the sun, the ratio of the major axes of the ellipses described by them were calculated with great exactness. A small number of observations judiciously combined sufficed to determine the other elements of the orbits, and each new observation confirmed the results obtained by previous observations and by the hypotheses upon which the calculations were based. The calculation of the perturbations in the movements of the planets

[1] For Venus 1′ 6″, for Jupiter 50″, and for Mars 26″ (at their least distances from the Earth).

produced by the other planets, in particular by Jupiter, led us to look upon the laws of Kepler simply as a first rough approximation, and supplied the very best possible *a posteriori* demonstration of Newton's law. Laplace's *Mécanique Céleste*, a brilliant synthesis of the work of the eighteenth-century astronomers, was the most perfect example of science which the human mind has been able to conceive. With the help of a very small number of hypotheses, namely the principles of mechanics and the law of Newton, celestial phenomena were now predicted with admirable accuracy, and the relative positions of the planets determined with perfect rigour.[1] Space was therefore conquered, thanks to mechanics. The only difficulty that remained was that of determining the absolute dimensions of the solar system; for all calculations would have remained the same, had the distance of the sun from the earth been multiplied by 0·99 or by 1·01; all distances would simply have been multiplied in the same proportion.

21. The Absolute Value of the Dimensions of the Solar System.

The methods enabling us to determine this essential element are very small in number, one of the most precise being based upon observation of the transit of Venus across the sun's disc, a phenomenon, however, that is unfortunately of very rare occurrence. If we watch this transit from two different points on the earth's surface, points the distance of which is well known, we determine triangles, the calculation of which enables us to know the angle under which the distance of the two points would be seen by an observer placed on the sun, and we can easily infer the distance of the sun from the distance of the two points. According to this method, then, the triangulation of the sky, if one may use the phrase, is

[1] The only real difficulty was the motion of the perihelion of Mercury, of which we shall speak again.

based upon the triangulation of the terrestrial globe. Thus astronomy, after rendering immense services to geodesy, becomes in its turn its debtor.

Another method for determining the absolute value of the dimensions of the solar system is based upon observation of the satellites of Jupiter. The method was originally proposed by Roemer as a means of solving the converse problem. Assuming the dimensions of the solar system to be known, Roemer set out to deduce from the observation of the eclipses of the satellites of Jupiter the value of the velocity of light. These eclipses are very frequent, on account of the vast dimensions of Jupiter, the proximity of the satellites, and the small inclination of their orbits; the nearest satellites are eclipsed at every revolution, that is to say every two (or four) days.[1] It is therefore quite easy to study these eclipses with accuracy and to discover the exact rhythm, though this varies slightly if we take into account the differences in the speed of Jupiter according to the position of the planet in its orbit. But according as Jupiter is in opposition or in conjunction, that is to say is to be found in the sky on the side opposite to or in the neighbourhood of the sun—the telescope enables us to observe this planet during the day, if it is not too near the sun—the distance from the earth to Jupiter varies by an amount almost equal to the diameter of the earth's orbit. The time required for light to traverse this distance is about sixteen minutes. Thus observation of the moment of eclipse, with an error not exceeding a second, makes it possible for us to determine this interval of about sixteen minutes to an accuracy of one thousandth part, that is to say with three exact figures. The distance, too, could be calculated with the same accuracy, were the velocity of light rigorously known. In the days of Roemer the science of physics was in far too backward a state to make it feasible to determine the velocity of light by a terrestrial

[1] More exactly, 42½ hours, and 3 days 13¼ hours.

experiment, and Roemer's observations furnished at once an approximate determination of this velocity, and the first positive demonstration that it existed at all. After Roemer, Delambre discussed very minutely, for the same purpose, the eclipses of the satellites of Jupiter. To-day, the velocity of light can be determined by numerous optical and electrical methods, all giving remarkably concordant results. It would thus be possible to deduce the dimensions of the solar system from calculation and the observation of the eclipses of the satellites of Jupiter. We thus see called in, for the determination of space, elements which, *a priori*, appear to be quite foreign to geometry. Not only mechanics but almost the entire science of physics is laid under contribution for the determination of the co-ordinates of the planets with reference to axes passing through the centre of the earth.

22. Galilean Axes.

As soon as knowledge of the real movements of the planets had led astronomers to substitute the heliocentric for the geocentric system of the ancients, it was of course natural that they should choose axes having their origin at the centre of the sun instead of that of the earth. More precisely, mechanical considerations induced them to take as the origin of the co-ordinate axes the centre of gravity of the solar system, a point which is not far from the centre of the sun (since the heaviest planet, Jupiter, has a mass which does not attain one-thousandth part of that of the sun). These axes are the axes of Galilei; we shall have occasion to emphasize their importance when dealing with the notion of time, a notion of which we have already made repeated use, and which it is now convenient to consider in detail.

23. The Sidereal Day.

It was the regular periodicity of days and nights which furnished man with the first natural clock, calling a clock any means enabling us to substitute accurate numerical determination for the vague idea of psychological time. This is not the place to give a history of the calendars, based upon the combination of the day (solar), of the month (lunation), and of the year (periodical return of the seasons; crossing of the meridian during the night by the same constellations), but it will not be out of place to recapitulate the definition of the sidereal day and of the mean solar day.

Independently of the natural clock formed by the sun-dial, man has for a long time been in the habit of measuring time by means of mechanical clocks. We are not going to give here the history of the progress of clockmaking, a history closely allied to that of the progress of mechanics. We shall accept as a fact the actual existence of clocks and of chronometers, the movements of which must be admitted to be regular, on account of the simple fact that they agree among themselves although based upon quite different mechanical processes (pendulum or spiral spring) and of very varied dimensions. This agreement may be compared to the agreement which we have already mentioned as existing between the lengths of bodies which we call solid, and from which we concluded that their dimensions were practically invariable. In the same way the mutual agreement within a few seconds of a large number of clocks and chronometers must be regarded as a proof of their regularity to the extent of that approximation. We may now assume that we possess one of these chronometers and are making use of it for the purpose of observing astronomical phenomena. If we measure the intervals of time separating two consecutive crossings of the meridian by the sun, we find, according to the season of the year, differences amounting to several seconds. But the differences we find between the intervals of time separating two consecutive passages of the meridian by any fixed star are, on the contrary, practically insignificant. More precisely, they do not obey any regular law,[1] and belong to the same order of magnitude as the relative differences between the movements of two well-regulated clocks. Experience has naturally led us to consider the *sidereal day*, that is to say the interval separating two consecutive crossings of the meridian by a star, as *rigorously invariable*; this hypothesis has always been confirmed by experience, the small differences found being regarded as due to the chronometers. It is therefore practically the observation of the stars which provides us with the means of regulating our chronometers. From the sidereal day thus rigorously determined it was easy to deduce the mean solar day, using the fact that a year contains one day more of sidereal than of solar days, on account of the superposition of the earth's rotation round the sun on its rotation about its own axis.

24. The Time of Astronomers.

Do there exist any theoretical reasons entitling us to consider the sidereal day as rigorously invariable? We must admit that there are good reasons for a negative answer, so that the constancy of the sidereal day must be looked upon as a first approximation only. It is, indeed, quite clear that the tides act as a sort of brake upon this immense fly-wheel which is the earth, a brake which must in the course of time bring about a retardation in the velocity of rotation, and consequently result in the prolongation of the day; the effect is, of course, very small, but it is not easy to evaluate its order of magnitude directly *a priori*. Does the use of chronometers furnish us with any sufficient reasons for regarding the effect as insignificant? Certainly not, because the science of chronometry is still

[1] We are ignoring here corrections to allow for aberration.

much too recent and imperfect to enable us to compare with certainty the indications of a chronometer existing to-day with those of a chronometer which was going, say, a century ago. We could try to obviate the difficulty by making use of the length of a pendulum beating seconds, but then we should have to assume arbitrarily that the value of g does not vary with time, and about that we evidently know nothing. Whence then comes the astronomers' belief in the almost absolute constancy of the sidereal day, and where is that mysterious clock entitling them to maintain that two intervals of time separated by, say, fifty years are rigorously equal? The answer to this question is very simple: the clock used by astronomers is none other than the whole solar system itself. The hypothesis of the constancy of the sidereal day has enabled us to specify a variable t, which we call time and which is thoroughly known, just as the axes of Galilei, having their origin at the centre of gravity of the solar system and directed towards the fixed stars, have furnished us with an accurate specification of space. *With respect to time and space thus specified, we may affirm the fundamental principles of mechanics*, and with the help of these and Newton's law we can calculate the position of the different bodies of the solar system at any future moment. Observation has enabled us in numberless instances to verify the accuracy of predictions thus made, and these verifications, supporting hypotheses in themselves simple, have given astronomers absolute confidence in these hypotheses, and consequently in the specification of time and space which they include. If, therefore, after the lapse of decades or of centuries, we find that all the phenomena are taking place a little sooner than they ought to take place according to the calculations, the difference existing between the calculations and the observations being always the same at a given epoch and increasing regularly with time, we shall conclude that a slight correction must be made in the measurement of time in order to make theory and observation entirely agree. In other words, we shall conclude that the sidereal day has grown slightly longer and that the interval of say 100,000 actual sidereal days separating two observations is in reality equal to 100,000 sidereal days of the year 1926 plus a fraction of a day. The phenomenon which should have taken place at the end of 100,000 sidereal days of 1926 will thus take place a little before the actual 100,000 sidereal days have elapsed; that is to say, before such and such a star has crossed the meridian for the 100,000th time. The observation will thus apparently precede the moment predicted by the calculation.

25. Privileged Axes and Privileged Chronology.

The laws of mechanics and astronomical observations thus lead us to adopt, for the measurement of space and time, privileged axes and a privileged chronology. Thanks to the choice of these axes and of that chronology, formulæ become very simple, whilst, on the contrary, with other axes and with a different chronology they would have been most complex. We shall pass over the somewhat trivial transformations called the transformations of Galilei, which consist in multiplying the number expressing time by a constant, and in substituting for the system of axes another system which with respect to the former is either at rest or has a uniform motion of translation. All space and time reference systems, differing only by the transformation of Galilei, are equivalent from the point of view of mechanics—they are equally privileged. For a more complete account of the facts we have just summarized we cannot do better than refer the reader to the luminous pages of M. Painlevé in the chapter devoted to mechanics in his work *Méthode dans les sciences*.[1] Between treating

[1] *De la méthode dans les sciences* (Nouvelle collection scientifique; Paris, Alcan).

certain axes and a certain chronology as privileged and treating them as absolute there is for many minds a difference merely of form, and there is little doubt that the majority of physicists and astronomers attached and still, perhaps, continue to attach an absolute value to the system of axes and to the chronology with respect to which the laws of mechanics are verified in a specially simple way. We shall see that the laws of relativity lead us to dispute this absolute character, and our readers will no doubt agree with us that it really is questionable. Is this, however, any reason for denying all value to the ordinary view? I do not think so. The discovery of the earth's sphericity and of the antipodes does not prevent the vertical from remaining a privileged direction in our daily life: it does not possess the absolute value which a man persuaded that the earth was flat would attach to it, but it nevertheless possesses a very great value, and a being who disregarded its privileged character would not live very long. If therefore we give up the use of the word *absolute*, a philosophical term which ought to be banished from scientific language, and replace it, as Poincaré invites us to do, *by a more modest epithet to which scientists will attach much the same meaning*, we may say that the choice of the axes and chronology of Galilei is particularly *convenient* for the physicist. But we must bear in mind that Poincaré considered it also as merely *convenient* to admit that the earth rotated round the sun and to believe in the existence of the world of perception. The doubt therefore which the employment of such a modest adjective as *convenient* seems to imply is a philosophical and not a scientific doubt. It is convenient in our daily life never to forget that the vertical direction is a privileged one; it prevents many mistakes, that, for instance, of a person living on the fifth floor of a house walking out through the window instead of through the door leading to the staircase. It is also *convenient* for us, in our work in mechanics or astronomy, to utilize privileged axes and a privileged chronology; it saves us from hopeless intricacy in our calculations, and enables us to solve many problems which would otherwise remain insoluble.

26. Are the Privileged Axes and Chronology Independent of the Earth?

In order to get as clear an idea as possible of the results obtained with the help of the axes and the chronology of Galilei, let us try to find out to what extent they may be regarded as independent of the fact that we are living upon the earth and not upon another planet. Could our definition of the axes and of the chronology be accurately communicated to beings resembling ourselves and living upon the planet Mars with whom we were able to communicate, either by wireless or by some other means?

In order to reply to this question we shall have to examine in turn the various elements of which the Galilean axes and chronology are composed.

In the first place, the origin of the absolute axes is the centre of gravity of the solar system; and it would appear, at first sight, that this definition is independent of the place where the observer finds himself, whether upon Mars or upon the earth. If, however, we reflect a little we find that the notion of the centre of gravity at a given moment presupposes an accurate knowledge of the position of the various stars *at the same moment*. We find here a difficulty to which we shall soon have to return.

The directions of the absolute axes are determined by those fixed stars which are far enough away to make the dimensions of the solar system absolutely insignificant in comparison with their distances; here we encounter no difficulty.

Finally, the unit of length chosen will be, for instance, the major axis of the earth's orbit. It seems again, at first sight, that this definition does not depend upon the place in which the observer is stationed, but we shall be less sure of this after we have discussed the theory of relativity. Let us proceed, however, and come to chronology; we must specify both the unit of time and the origin of time. In order to specify the unit of time we can choose a *celestial clock* visible both from the earth and from Mars, such as, for instance, the period of revolution of one of the satellites of Jupiter. The exact observation of this period presents great difficulties, practically insurmountable; but these will diminish if, instead of one revolution, we consider a great number of revolutions, say 1000, 10,000, or 100,000. The inevitable errors will then be divided by the number of the revolutions, and we may hope to obtain an accuracy comparable to that attainable in ordinary observations. As for the *origin* of time, the difficulties are much greater, but why not obviate them by allowing a certain arbitrariness; the introduction of a small parasitic constant will in no way modify the equations.

27. Introduction of the Velocity of Light Necessary.

The question which now arises is whether the image which a Martian observer has formed of the solar system will be exactly the same as that formed by an observer upon earth. We must consider this: when we are making an astronomical observation, we watch for the coincidence of the centre or of the edge of a star with a fine thread placed in the reticle of a telescope, and we note the moment of this coincidence. It is from these data, including those by which the position of the telescope is fixed, that we are able to infer the position of the star. Now it is clear that the ray of light has required a certain time to reach us. If the ray of light has had to travel 15 minutes before reaching us, then what we are observing and specifying is the position of the star not at the moment indicated on the clock but 15 minutes earlier. We have here a correction necessary for all astronomical observation. Now, in order to ascertain the time required for the light from the star to reach us, we must know two elements: the distance of the star, and the velocity of light. If the distance of the star is, say, 270 million kilometres, and the velocity of light is 300,000 kilometres per second, we find that it has taken the light exactly 900 seconds, or 15 minutes, to reach us. Our observations give us the position of the star at the hour indicated by the clock *minus* 15 minutes.

Thus all our knowledge of planetary space presupposes, in addition to the hypotheses already set forth, the essential hypothesis that the velocity of light in empty space is known to us, that is to say, that it is constant. If the postulate of the constancy of the velocity of light *in vacuo* were to fail us, we should have to give up the hope of ever knowing the position of the stars at a given moment indicated on our clocks (unless, of course, it were possible for us to substitute for this missing postulate another, differing from it but little, and introducing only small corrective terms; but even in this case our knowledge would lose much of its simplicity). We are thus compelled, for the complete definition of time and space required by celestial mechanics, to call to our aid the velocity of light and its constancy. And here the fact that the stars observed are in motion relative to each other, introduces an essential difficulty, which lies at the very core of the new theories of relativity. We shall discuss this difficulty in detail in Chapter VI. We may, however, say at once that the corrections introduced by the theory of relativity are small, and it may not, therefore, be quite profitless to sum up the results which one may consider as established, approximately at least, with respect to knowledge of space and time.

28. Approximate Results retain Scientific Value.

From the preceding discussion it follows that these results are only established to a certain degree of approximation, but we must not attach too great importance to this remark, or take the view that the results, because they are approximate, are simply wrong, forgetting altogether the fact that they are very nearly right. This would be to adopt the uncompromising attitude, which, apparently at least, was sometimes that of Poincaré, but against which, in my opinion, one cannot protest too strongly. From the purely mathematical point of view there is no doubt that the equation $a = b$ is equally false, whether the difference $a - b$ be one of a thousand-millionth or of a tenth (supposing for definiteness that a and b are comparable with unity); and if we have a false equation there is nothing easier than to deduce from it, by means of algebra, any other false equation; there are no degrees in error. But from the point of view of the physicist, such a thesis is untenable; all measurements in physics being only approximate, must we conclude that they are all equally false and that we ought not to attach any more value to observations furnishing us with seven exact decimals than to observations which are grossly erroneous, and fail to discriminate between a quantity and its double or its half? No one would dare uphold such a paradox. The whole history of physical theories goes to teach us that when a simple law, at first believed to be exact, is reduced, in consequence of more accurate measurements, to the rank of only an approximate law, it nevertheless retains a great part of its interest. Such is the case with Mariotte's law, for instance; and even more instructive examples are furnished by the facts that atomic weights are proportional to simple whole numbers, and that the cleavage planes of crystals can also be brought into relation with simple integers. We know now that these laws are only approximate, but they are not entirely false for all that, in the sense that they still contain a part of the truth, and only need to be completed. This remark will no doubt also apply to the Galileian space and chronology, and although we know that they give a too simplified image of a more complicated reality, we need not deny all value to this image. It enables all dwellers on the earth to agree upon a definition of time, thanks to which we can calculate the position of the planets in space, with respect to the Galilean axes, with a precision surpassing even that of the observations themselves. More than a century has elapsed since Laplace wrote his book on celestial mechanics, and only one phenomenon, the motion of the perihelion of Mercury, has escaped the close network of deductions, the ensemble of which constitutes the Galilean notion of time and space. We shall see that the theory of relativity gives an account of this hitherto unexplained phenomenon, but this alone would not have been a reason sufficient to make scientists adopt the theory. It is only one reason in addition to many others. This motion of the perihelion of Mercury is very slow, the residue unaccounted for amounting only to forty-three seconds of arc per century. It is one of those infinitely small residues which turn the scale of theories and amply justify the often derided care with which numerous calculators and observers have devoted years to the accumulation of numbers, which will perhaps never serve any purpose, but which, on the other hand, may lead to a scientific revolution increasing man's means of activity tenfold.

29. What do we know of Interstellar Space?

Let us now pass beyond the confines of the solar system and try to state in precise form the data we possess on stellar space. We have been able to obtain the distances of a certain number of the stars nearest to us by means of a sort of triangulation, taking as the base of the triangle the diameter of the earth's orbit, and determining the angle at the top of this triangle. This means that we are estimating the apparent displacement of a given star in relation to other more remote stars, when we observe it on two occasions at an interval of about six months, the earth then occupying opposite ends of the same diameter of its orbit. Other very ingenious methods, based upon interference effects, have been recently used for the purpose of estimating the distances and the dimensions of the stars. All these operations are naturally based upon optics, since all that reaches us from the stars is the light they emit. We are unable to derive any information with regard to the properties of space, since these very properties are postulated in our construction of interstellar space. Take for instance two stars, the light from which reaches us only after twenty years, the angle between the rays by which we see them being 60°; we conclude that the two stars and the earth are the three vertices of a gigantic equilateral triangle, and that consequently light ought to take twenty years from one star to the other also. We have, however, no direct proof of this conclusion, and we cannot conceive at present how the conclusion could be tested even in the roughest way by direct observation. We shall revert to this point in Chapter V.

CHAPTER III

Abstract Geometry and Geographical Maps

30. The Abstract Conception of Geometry.

In the preceding chapters we have considered geometry as a physical science, and turned to experience in order to define space. This point of view seems to us to be the only one from which we can learn anything with regard to perceptible reality. It is, however, impossible to leave entirely unmentioned the abstract conception of geometry, a conception which, if necessary, could easily be justified by the services it has rendered and still continues to render to mathematicians. Apart from its utility, abstract geometry possesses a beauty of its own which assures it of cultivators as long as the æsthetic taste for pure science continues to exist. It helps, moreover, to develop and to uphold this taste, the disappearance of which would be the greatest calamity that could befall humanity. The day when pure science ceases to be cultivated for its own sake and when its peculiar beauty ceases to appeal to the imagination of some at least of the choicer spirits—that day will speedily be followed by the decay of material civilization. Another few centuries, and humanity would return to primitive savagery, or, even worse, would evolve towards an organized and stereotyped barbarism, comparable to the monotonous life of ants and bees.

But however great may be the interest of the abstract

conception of geometry, it would nevertheless, if absolutely necessary, have been possible without much inconvenience to leave it undiscussed in the pages of a book concerned with the study of space as a reality. On the one hand, however, I thought that some readers would have been disappointed by the omission, whilst, on the other, certain explanations which we shall have to give on the subject will be useful as an aid to the comprehension of general relativity (Chapter VII). The present chapter should, however, be considered as a sort of parenthesis, and the reader who is not attracted by the discussion, which, as is inevitable in this subject, is occasionally somewhat abstract, may with little inconvenience omit it.

Poincaré once wrote that " there was no sense in saying that Euclidean geometry was either true or false ". This remark may be compared with the famous phrase of an English philosopher who is also a humorist: " Mathematics is a science in which you never know what you are talking about and never care whether what you are saying be true or false."

31. A few Remarks on Mathematics.

I have no intention of discussing here the meaning and the value of mathematics as an independent science. Some widely current sophisms would have to be refuted, in particular the one according to which the science of mathematics never creates anything, but contents itself with transforming elements coming from outside. It were as reasonable to say that a beautiful poem was nothing but a combination of letters and that the picture of a master was only a heap of colours. Only the typefounder and the colour manufacturer have created anything; both the poet and the painter have merely arranged in a certain manner elements furnished by others, and have created nothing. But let us confine ourselves to geometry. What we have been studying hitherto is the science of real space, using the word real in the same sense as when we say that physics deals with real objects and phenomena, the metaphysical discussion as to what is real being outside the scope of science. Now for Poincaré geometry is a purely abstract science, an ensemble of deductions derived from a certain number of axioms and postulates. It is clear that such an ensemble of deductions can neither be true nor false, since in any particular case all we assert is that if the axioms and postulates be true, then such and such consequences are equally true. In these axioms and postulates there figure words which are defined solely by these very axioms and postulates and to which we are consequently not obliged to attribute any concrete meaning; and these same words will also figure in the deductions. It is thus literally exact to say not only that we never know whether these deductions are true or false, but also that we never know what we are talking about when formulating these deductions, since words appearing in them have no concrete meaning *a priori*. It is with geometry thus understood that we are first going to deal as briefly as possible.

32. Analytical Geometry a Means of Defining Geometrical Conceptions.

The simplest and most elementary way to present this conception of geometry is, as we have already said, to start with analytical geometry. Let us take three rectangular axes; for the moment these words have no other meaning than the following: Our axes are *things* which we shall call Ox, Oy, and Oz respectively, the letter O indicating the origin of the axes, which, by definition, has as co-ordinates $x = 0$, $y = 0$, $z = 0$; for all points of the axis Ox the co-ordinates y and z are 0, and the co-ordinate x may have any value; and similarly for the other axes. A point in space will, by definition, be the aggregate of any three numerical values of x, y, and z, as, for instance, $x = 2$, $y = 3$, $z = 0{\cdot}52$. A plane is the aggregate of points whose co-ordinates satisfy an equation of the first degree, as, for instance, the following: $2x + 3y + 5z - 12 = 0$. A straight line is the aggregate of points common to two planes. The angle between two planes or two straight lines is *defined* by the formulæ which, in the ordinary method of treating analytical geometry, we *prove*, the proofs being based on the results of ordinary geometry and trigonometry, which are supposed to be known. These definitions being granted, we can easily prove that the axes of co-ordinates are straight lines at right angles in pairs. We shall define the distance between two points by means of the classical formula in analytical geometry. We can then demonstrate step by step all the theorems of elementary geometry. These theorems once demonstrated in the abstract, we can give them a concrete, or rather schematic, representation by drawing the figures of Euclid's geometry. We shall see in a moment what value we can attach to such a schematic representation. First, however, I shall glance briefly at an important property of figures in concrete geometry, a property which does not present itself naturally in abstract geometry; it is the notion of *sense* in figures.

33. The Notion of Sense—it is Incommunicable.

This idea of *sense* is one of the most familiar ideas of civilized man, even when he knows nothing of geometry, as it may in the end be reduced to the distinction existing between our right hand and our left. The right hand and the left of a normally constituted man are alike,[1] so that a glove-maker, if he possesses the measurement or a cast of one hand, will have no difficulty whatever in making a pair of gloves which will fit both hands exactly, on condition, of course, that the right glove is put on the right

[1] There are differences in the lines of the hand and, in particular, in the finger-prints; these differences may be of interest to chiromancers and to police anthropometrical departments, but they need not detain us here.

hand and the left glove on the left hand. What we call *sense* in geometry is the character which distinguishes the right-hand from the left-hand glove, and vice versa. We may try to define this character by saying that one glove is the image of the other, as seen reflected in a mirror, but this definition is incomplete, for it does not tell us *which* of the two gloves is the right and which is the left. These examples of the hand and the glove are simple because they are so familiar, but geometrically they are quite complicated. We could give other examples, scarcely less familiar, and geometrically more simple. Take, for instance, a pair of ordinary scissors. We usually hold these with the thumb of the right hand in one ring and the forefinger in the other.[1] When we close the scissors the natural movement of the hand tends to press the two blades against each other, so that, even if there is a little play in the screw which keeps them together, they will cut a thin sheet of paper neatly without spoiling it. If we take the same pair of scissors in the left hand, the natural movement of the hand will, on the contrary, tend to separate the two blades when we close the scissors, so that unless the screw is very tight, and the paper kept stretched, it will be quite impossible to cut it. For a left-handed man we should have to make a special pair of scissors which would be the image reflected in a mirror of an ordinary pair as used by a right-handed man. It is obvious in fact that the image reflected in a mirror of a workman using a pair of scissors with his right hand is simply a picture of a workman using a pair of scissors with his left hand.

If we now open a pair of ordinary scissors and place them on a table, one of the blades will be resting on the table, and we shall see its flat side, whilst the other blade will be above the first, showing us its back, more or less

[1] In a large kind of scissors used by tailors, the rings are unequal and irregular, one being for the thumb and the other for the forefinger and middle finger together.

convex; the first blade may be called the lower and the other the upper. We could also turn over the scissors so that the lower blade will become the upper and vice versa. With certain scissors we could not tell that any change had been made, unless we first made some mark on one of the two blades, these being identical.[1] Anyhow, in whatever position the scissors are placed on the table, in order to close them we must make the point of the upper blade turn in the same direction as the hands of a clock; the direction would be reversed in the case of scissors made for a left-handed person. If the scissors be opened on the table at a right angle, we can call the upper blade the axis of x, the lower blade the axis of y, and the vertical directed upwards the axis of z, these three axes constituting a trirectangular trihedron, the sense of which will be called right or direct in the case of ordinary scissors, and left or inverted in the case of scissors made for left-handed persons.[2] This is the simplest geometrical diagram to which we can attribute a *sense*.[3]

34. The Notion of Sense.

If we start from a determined system of Cartesian co-ordinates, there is thus no difficulty in defining the sense of the figures formed by points given by their co-ordinates. We generally reduce this definition of sense to the calcu-

[1] There is usually, however, a difference between the two ends of the pin which connects the two parts, one end only being provided with a nick for adjusting the tightness of the screw.

[2] We can obtain a representation of such a trihedron by opening the right hand and bending forward the middle finger, which we may take as axis of x, the forefinger being the axis of y, and the thumb the axis of z.

[3] Among common objects showing sense, we may mention also a kind of coffee-pot called a "pourer" which is held by a long handle at right angles to the spout or nozzle. According to the sense of the trihedron formed by the vertical axis of the coffee-pot (directed upwards), the handle, and the spout, it is more convenient to work with it in the right hand or in the left; a person using two pourers, one in each hand, would find it a great advantage to have them of different senses (images of each other in a mirror).

lation of a certain number which is defined by means of the co-ordinates, and is positive or negative, according as the sense is right or left. This, however, essentially presupposes a knowledge of the *sense* of the trihedron taken as the trihedron of co-ordinates, and analytical geometry cannot furnish us with such a knowledge. Generally speaking, no operation of an analytical or algebraical kind would enable us to distinguish between a right and a left trihedron, or, if we like, between the right hand and the left. The only way to do so would be to *show* them, that is to say, to have recourse to the visual intuition of space, supplemented by the natural distinction which from our childhood we have been in the habit of making between our right hand and our left. Had education[1] not inculcated in us this sense of asymmetry, which, by the way, is also forced upon our attention by certain manufactured articles (such as clocks, keys, printing type, books, &c.), it is quite probable that two objects, images of each other in a mirror, would be in a sense indistinguishable for us, that is to say, we would notice they were different without being able to tell *which* was right and which was left. We know that there are children who when learning to write sometimes trace characters which are the reflection in a mirror of the ordinary letters, without noticing that they are not the same.

In the temperate zone the motion of the sun furnishes us with a means of distinguishing between the right hand and the left; such is not the case in the torrid zone, where careful observation of the stars would be needed before a man who does not know already could tell which hand is his right and which his left. Of course any well-known landscape could be utilized for the same purpose, but

[1] Though I use the word " education " I have no thought of settling the question of the part played by heredity in the fact that man is right-handed rather than left-handed or ambidextrous; we may, if we please, understand education in a very wide sense, including education of the race as well as of the individual.

people living in forests and not able to see the stars would certainly experience a real difficulty in defining their right hands (unless they were to make use of the asymmetry of their own bodies). I have dwelt a little upon this difficulty because it very often passes unnoticed, the distinction between our right hand and our left being so familiar to us from our very infancy. If we—to assume the impossible —ever forgot the distinction, we would constantly be reminded of it by printed writing or manuscript, by the movement of the hands of clocks, &c.; but these are acquired notions impossible to be transmitted except by sight and example. If we were able to communicate, by wireless telegraphy, with the inhabitants of some planet covered by clouds so that no stars could be seen, it would be extremely difficult for us, if not absolutely impossible, to make sure whether they were calling right the hand we are calling right, and left the one we are calling left. No doubt, we could manage it if these inhabitants could repeat the experiments made by Pasteur which proved that certain minute forms of life do not react in the same manner towards right-handed crystals as they do towards left-handed ones. But how can we be sure that these extremely small creatures exist upon another planet, and that they are identical with those we know and not with the reflection of these in a mirror?

35. The Euclidean Schema.

On the other hand, all geometrical properties except that of *sense* can be expressed analytically and are, therefore, easily transmissible to beings who are endowed with an intelligence analogous to our own and with whom we could communicate by means of signals. These analytical properties can be made perceptible by the use of geometrical schemas; these schemas once defined, it is clear that we could abridge certain analytical demonstrations, substituting for certain groups of equations corresponding geometrical expressions. For instance, once we have established that a sphere and a straight line have two points in common, except in the case where the straight line is a tangent to the sphere, in which case it is perpendicular to the radius and the two common points coincide, we can reason about a sphere and its tangents without having to rewrite the equations from which we derived these properties. We can even draw a figure, or more exactly a schema, enabling us to follow the reasoning more easily, should it become a little complicated. The simplest schema which we can bring into correspondence with analytical geometry with three variables is the *Euclidean schema*; and in this sense Euclidean geometry is true, if simplicity be taken as a criterion of truth. This truth, however, is only a relative one, for instead of the formulæ of Cartesian geometry we can imagine many other systems of analytical formulæ, and the simplest schemas corresponding to these systems of formulæ are furnished by the geometry of Lobatchewski, or by those of Cayley, or by that of Riemann. These questions are quite distinct from that of the experimental truth of one or the other of these geometries.

36. Example of a Schema of Imaginary Geometrical Elements.

It will help to give us a clear idea of what is meant by the schema, if we note that it may be applied without modification to problems involving imaginary elements or elements situated at infinity. Let us take an example. It is easily proved that a plane curve of the third degree (or cubic) possesses nine points of inflexion, and that every straight line joining two of these points of inflexion contains a third. We easily conclude that there exist twelve straight lines, each of these containing three points of inflexion, each of these points being situated upon four of the straight lines. The configuration thus defined presents the peculiarity that it cannot be realized by means of elements that are all real, as a cubic cannot have more than three real points of inflexion, which are in a straight line. It is, however, possible to trace a schema (fig. 2) upon which the points of inflexion are indicated by the numbers 1, 2, 3, 4, 5, 6, 7, 8, 9, the twelve straight lines joining them three by three being as follows:

123
456
789

147
258
369

159
627 curve
384 curve

429 curve
357
186 curve.

We have had to represent four of the straight lines by lines visibly curved; this is a necessary imperfection of our schema, whilst another inevitable imperfection is the following: The points marked A, B, C, D, E, F in our diagram must not be considered as actual points intersecting the lines crossing each other; the point E, for instance, is found both on the *straight* lines 186 and 726, but these straight lines, crossing one another already in

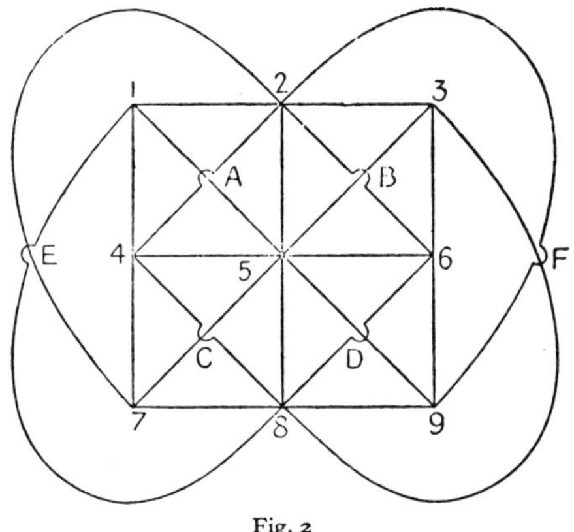

Fig. 2

point 6, cannot have any other point in common. We must thus imagine our schema to be viewed as it were in perspective, so that at the points A, B, C, D, E, F the lines pass one behind the other without intersecting.[1] Under these conditions the schema gives us in a geometrical form the properties of the numbers (complex), which represent the points of inflexion of a plane cubic. It matters little so far as concerns the clearness of this representation that the straight line 627 is drawn as a curve.

37. The Schema of Spherical Geometry—Riemann.

Once the idea of the schema is understood, it is easy to generalize it. For the sake of clearness and simplicity in the exposition, let us confine ourselves to two-dimensional geometry; nothing essential would be changed if we were to pass to three or a greater number of dimensions. Cartesian co-ordinates can be represented upon the plane by means of a rectangular network. If we draw all the straight lines parallel to the axis of x at intervals of, say, one-tenth of a millimetre, and do this also for the axis of y, it will be easy to number these straight lines. The co-ordinates of any point will simply be the numbers of the straight lines passing through that point. The conception of co-ordinates thus understood can be generalized on any surface; for instance, on the surface of the earth, without assuming it to be spherical, we can number the meridians and the parallels by giving an astronomical definition of both latitude and longitude. There is then nothing to hinder us from considering the points which have the same co-ordinates as corresponding to one another on the plane and on the sphere. It is upon this principle that the con-

[1] If we were anxious for completeness, it would be necessary to show in our schema the points of intersection of lines such as 123, 456, 789, and the three analogous groups; each of these four groups of three lines forms an actual triangle. But our object was to give an example and not to build up a complete theory of the diagram of the points of inflexion of a plane cubic.

struction of geographical maps is based. It is clear that if we map a portion of the sphere upon a plane, and apply the terms straight line, circle, triangle, &c., on the sphere to all the elements corresponding point for point to the elements of the same name on the plane, all the theorems of geometry which are true on the plane will be equally true on the sphere—provided we constantly bear in mind the conventional meaning of the terms appearing in the enunciations of the theorems. Nevertheless, if this meaning is too far removed from reality and simplicity, then the enunciations will be purely verbal, and without interest.

Riemann's geometry consists in calling straight lines the great circles traced upon the sphere. If we consider only a portion of the sphere less than a hemisphere, then two straight lines cross one another only in one point, otherwise they have two points in common. The sum of the angles of a triangle is always greater than two right angles; from this theorem follows an important consequence with respect to parallelism and the conservation of direction, a question to which we shall have to return (see § 41).

38. **Plane Schema for Any Geometry.**

Instead of representing, as it were, plane geometry upon any surface, such as the sphere, we can proceed in the converse way and represent the geometry of any surface upon a plane. This is what Poincaré did in the theory of automorphic functions (or Fuchsian functions). He considers the triangles formed by arcs of circles, the centres of which are situated upon the same straight line (which he takes for the axis of x); by definition, the sides of these triangles are "straight lines". Confining himself to the consideration of a portion of the plane situated above the axis of x, he shows that through two points one straight line, and only one, can be drawn; but that through a given point an infinite number of straight lines can be drawn which never meet a given straight line; they are included within the angle formed by the straight lines intersecting the given straight line on the axis of x. Angles are defined as in ordinary geometry, and the distance between two points as the logarithm of the anharmonic ratio which they form with the two points in which the straight line joining them intersects the axis of x. Under these conditions, the theorems of the hyperbolic geometry of Lobatchewski hold good—a remark which was found very useful by Poincaré in his researches on automorphic functions.

39. Well-known Examples of a Schema.

A well-known example of a geometrical schema serving the purpose at least as well as an exact representation occurs in the distorted maps we often come across in railway guide books. The railway lines are often shown displaced in a more or less arbitrary way so as to make the map easier to read, the sole object of the map being to show as clearly as possible the relative positions of the junctions and the various lines running into them. In a similar way some ancient geographical maps give one the impression of being schemas rather than exact representations. Even our own maps, when they deal with vast tracts of the earth's surface, such as the continents of Africa and America, are bound to be only schemas, the curvature of the earth preventing us from preserving the exact proportions of all the parts of large surfaces.

40. Mercator's Projection.

One of the most instructive ways of constructing a geographical map is to use *Mercator's projection.* The network of parallels and meridians is in this case represented by straight lines dividing the plane into a network of rectangles, in such a way as to preserve the similitude of corresponding infinitely small elements, so that the map of any small portion of the earth's surface shows the true shape without distortion. Let us try to get a clear idea of how this result can be obtained.

Consider two neighbouring meridians PA and PB (fig. 3), intersecting the equator in two points A and B distant say 100 kilometres from each other. We shall have to represent them by two straight lines perpendicular to the straight line representing the equator, such as *ace* and *bdf* (fig. 4). Let us take from the equator on these meridians

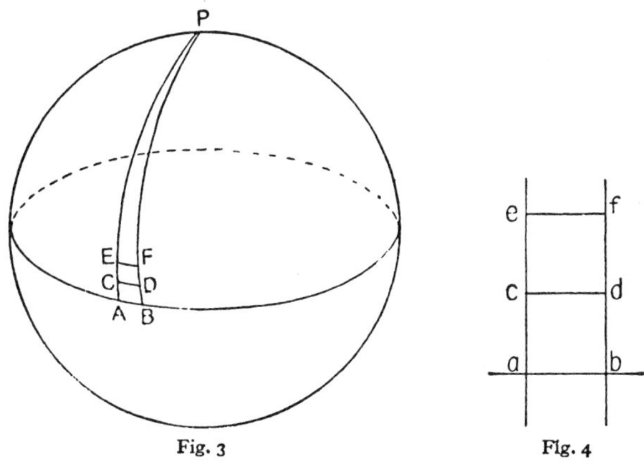

Fig. 3 Fig. 4

the lengths AC and BD, equal to the arc AB, that is to say, 100 kilometres; the curvilinear quatrilateral ABCD will be approximately a square which we represent on the plane by the square *abdc*. There is, of course, a small error, CD not being equal to AB on the sphere; the nearer to each other the meridians PA and PB are taken, the smaller the relative error will be. The error, however, must not increase, when we consider a new quasi-square on the sphere CDFE for which CE and DF are equal to CD; this quasi-square will be represented by a square *cdfe* which is necessarily made equal to the square *abdc*, that is to say *ce* is equal to *ac*, whilst CE is smaller than AC.

When we approach the pole, the two meridians of the sphere will be almost indistinguishable from straight lines, say MP, NP (fig. 5); to a quasi-square MNN′M′ corresponds a square *mnn′m′* (fig. 6), always equal to the primitive square *abdc*. If the angle at P is such that MN is equal, for instance, to 1/100 of PM, and if we take, as we have indicated, MM′ = MN, we see that PM′ will be equal to 99/100 of PM;[1] PM″ will also be equal to 99/100 of PM′. We see that we can continue indefinitely *without ever reaching the point P*, since every time the distance simply becomes 99/100 of the preceding distance, whilst the corresponding distances *mm′*, *m′m″*, &c., are always equal to the constant distance of the parallels on the plane, a square being made to correspond to a quasi-square. To obtain Mercator's projection rigorously we must diminish the angle at P towards zero; our quasi-squares then become, in the language of analysis, infinitely small squares, and there is perfect similarity between the infinitely small elements. The principal conclusion, however, remains: When the point M approaches the pole, the ratio between the areas of the plane and the spherical figure increases indefinitely, and the point *m* recedes to an indefinitely great distance.

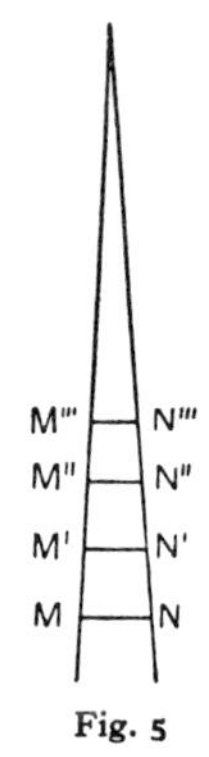

Fig. 5

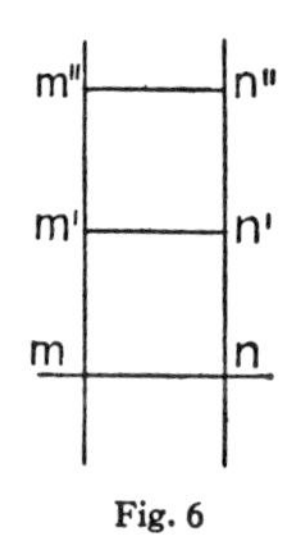

Fig. 6

41. Applicable Surfaces and Parallelism.

In the next chapter we shall study the difficulties arising from the fact that Mercator's projection extends to infinity in this way, although the sphere to be represented is finite. We shall see that these difficulties are not casual, but

[1] For the sake of clearness in the figure we have had to take the angle greater than this.

essential. It is impossible to avoid them, even by arbitrary modifications of the customary conventions with respect to the measurement of length, or, in other words, by giving up the physical postulate of the quasi-invariability of solid bodies. In concluding this chapter we shall, on the contrary, provisionally admit this postulate and briefly indicate the consequences that may be drawn from it with respect to the properties of lines traced on surfaces. The study of these properties originates with the famous memoir of Gauss, *Disquisitiones generales circa superficies curvas.* They are explained with incomparable mastery in the beautiful work, *Leçons sur la théorie des surfaces*, in which Gaston Darboux has given an account in didactic form of the researches of the nineteenth-century geometers, among them his own. We shall say that two surfaces are *applicable* to each other when we are able to establish a correspondence between them, point by point, in such a manner that lengths and angles are preserved unchanged. This definition presupposes that the distance between two points of a surface has been defined. We could avoid this difficulty by confining ourselves to the consideration of infinitely small distances. Such, indeed, is the logical course from an analytical point of view. But we may equally well agree that this distance is to be, by definition, the shortest distance between the two points under consideration. In practice, this shortest distance can be obtained by means of a thread stretched along the surface.[1] The shortest line thus indicated by means of a stretched thread is called a geodesic line of the surface; on a plane, the geodesic lines, or geodesics, are straight lines. It is clear from the definitions that on two surfaces which are applicable to each other, the geodesic lines on either surface will correspond to those on the other.

[1] For the thread to be stretched, it must be placed on that side of the surface which is convex for the line considered: if the points are too far away from one another, we shall have to take one or more suitably chosen intermediate points as relays.

Two surfaces are not in general applicable to each other, and, in particular, a surface is not in general applicable to a plane. A sphere is not applicable to a plane. We can see this very easily by noticing that it is easy to transfer to a surface applicable to a plane the notions of direction and of parallelism which are familiar to us upon the plane. Let us consider, for the purpose of fixing our ideas, a portion of the surface of the earth so small that it may be regarded as plane. Let us assume the directions of the cardinal points to be known, and let us fix all other directions by reference to these. If we now place ourselves upon a geodesic, that is to say upon a straight line, our direction with respect to the cardinal points will always remain the same. If we were to leave this straight line and follow a path along another straight line, the new direction would make with the first an angle which can be measured at the point of intersection of the two straight lines. If we describe a polygonal circuit, we shall return to the primitive direction after having turned altogether through 360°. This notion of the invariance of direction, no matter what closed path we traverse on the plane, is tantamount to Euclid's postulate.

It is easy to convince oneself that things do not happen in that way in the case of a sphere. If we change our position upon the earth, following a great circle, we shall be changing our direction relative to the cardinal points (except in the particular case where we are following a meridian, and in the even more special case in which we are following the equator). The result is that upon a sphere it is impossible to define parallelism as upon a plane, by merely assuming that we do not change direction when we walk straight on by the shortest route, that is to say, when we follow a geodesic.

42. Geodesic Lines and the Invariance of Direction.

Would it be possible for very small beings, imagined to be living upon a surface, to arrive at the notion of a geodesic without measuring lengths or stretching a thread, but simply by trying to walk straight along, without changing direction? The answer is in the affirmative, and depends upon an important geometrical property of geodesics considered as shortest lines. This property consists in the fact that their osculating plane is normal to the surface. In ordinary language we can express this in the following way: Let us imagine that certain very small beings living upon the surface plant vertical surveying-staves at various points, so that the staves are perpendicular to the surface. Generally, it would be impossible for them to put the staves in a line, that is to say, so that several of them could be seen in the same plane, viz. the plane determined by one of them and by the eye of the observer. If, however, they made the alignment as exact as they could, carrying on the process step by step, the staves so aligned would mark out a geodesic. On this point calculations confirm the conclusion of common sense, that we must always walk straight on and never take a roundabout road, that is to say, never alter our direction, if we wish to follow the shortest route.

We are thus entitled to define geodesics as being lines such that when we follow them we are always walking in the same direction. It is essential, however, to understand that on the basis of this definition it is not possible to define parallel directions at two or more different points of a surface. To put the matter precisely, if we try to preserve our sense of orientation during the process of describing a closed path on the surface, we shall find, when we return to the point we started from, an error in our estimate of direction which will be the greater, the larger the portion of the surface which the path encloses.

On the terrestrial sphere, for instance, we might describe a very small square (with sides, let us say, 1 kilometre long) and return to the point we started from after turning four times through a right angle. The error of direction thus made would be nil, if the earth were plane, and is in fact practically insignificant, on account of the vast dimensions of the earth in proportion to 1 kilometre. On the other hand, if we tried to describe a square of side 10,000 kilometres, we would notice that we had returned to the point we started from after describing three sides, that is to say, we would discover that a triangle existed having each of its three sides 10,000 kilometres and its three angles all right angles. We can obtain such a triangle, for instance, by taking the meridians corresponding to longitudes 0° and 90° E., from the north pole to the equator, and the arc of the equator joining the two points of longitude 0° and 90° E. If we describe the perimeter of this triangle, we shall come back to the starting-point, following the direction taken at the start, after turning three times through a right angle, that is to say *three* right angles *in all*, instead of *four*. If we were under the impression that it was a plane surface upon which we were moving, we would therefore be making an error of a right angle in our estimate of direction.[1]

43. **Utilization of the Linear Element.**

In order to define the intrinsic properties of lines traced upon a surface, properties which are the same for all applicable surfaces, there is no need to know the surface itself, but only the analytical expression to which is given the name *linear element*.

We shall write out this expression explicitly in Chapter VII. For the present it is sufficient, without writing down

[1] For the analytical discussion of these questions, in the general case when the contours considered are not necessarily geodesics, reference may be made to Vol. III of Darboux's *Leçons sur la théorie des surfaces*.

any formula, to state that a knowledge of the linear element enables us to solve any problem relating to distances and angles upon a surface. Conversely, if we can survey a surface, that is to say, measure empirically the shortest distance between any two points on it, the linear element will be known. The *a priori* introduction of the linear element is thus based upon certain hypotheses, which were brought to light by Riemann in a famous memoir, the most important of these, as it seems to me, being one which may be called the postulate of the ellipsoid, to which we shall return in the next chapter (§ 47) and in Note II.

CHAPTER IV

Continuity and Topology

44. The Very Small More Difficult to Reach than the Very Great.

Up to this point we have studied the properties of space and time chiefly on the human scale. It may now be interesting to inquire whether the properties found to be exact on this scale are equally so on a much larger or on a much smaller scale. The infinitely small and the infinitely great are, of course, alike inaccessible to us, and speculation concerning the infinite belongs to the domain of philosophy rather than to that of science. But there is no reason why we should not ask whether it may not be possible to take a few steps upon that infinite path, the ends of which are beyond our reach. The difficulties we encounter in studying the very small are not less formidable than for the very great. To some extent they are even more irritating, as there is in us a more or less unconscious feeling that the minute particles which compose solid bodies, elusive though they be, are in a sense within our grasp, whilst we willingly resign ourselves to consider as inaccessible a nebula the light from which would take a million years to reach us.

In reality, however, the difficulties, although of a different nature, are not less. We can even say, without paradox, that if we compare distances on the human scale, we can advance farther in the infinitely great than in the infinitely small. Light travels 300 million metres per second, which we can write as $3 \cdot 10^8$; the day contains about 100,000 seconds, that is to say, 10^5; and it takes more than 1000 days for light to reach us from the nearest fixed star. We have here a distance the order of magnitude of which we know; and, at this distance, a star many of whose properties spectrum analysis has revealed to us, the proper motion of which we know, and so on. Now this distance belongs to the order of 10^{16} metres, so that we require 16 successive multiplications by 10 to reach it, starting from ordinary sizes. If, however, instead of 16 successive multiplications by 10 we were to make 16 successive divisions, we would come in the end to dimensions belonging to the order of 10^{-16} metres or 10^{-14} centimetres. Since the linear dimensions of molecules are considered to belong to the order of 10^{-8} centimetres, we are now dealing with dimensions which are a million times smaller than the linear dimensions of the molecules. Even the magnificent and daring speculations of Bohr [1] have not yet enabled us to go so far. In the sky we go thousands of times farther, as the above calculation refers to the star nearest to us.[2]

What, we may ask, are the properties of space and matter on this minute scale? We cannot say, so that we may quite properly ask whether our intuition may not be deceiving us when we assume that the geometry of similar figures is applicable to extremely small dimensions. Anyhow, there is no doubt that there are certain results which, though evident according to the arithmetical conception of space, appear paradoxical to anyone who considers them by geometrical intuition. Here is an actual classic example from the theory of functions in mathematics.[3] I will explain it in quite an elementary way.

[1] See Jean Perrin, *Les Atomes*.

[2] We may point out in passing the use of optical properties for measuring very small lengths; we shall deal with this in Chapter V.

[3] See my *Leçons sur la théorie des fonctions* (Gauthier-Villars). It was the elaborate study of this example which led me to the theory of the measure of aggregates of points, a theory which has had a far-reaching influence on the theory of functions of real variables.

45. Geometrical Intuition at Fault in the Infinitely Small.

Consider a straight line 1 metre long. We may think of it as an extremely thin material bar. We naturally assume the matter to be continuous and infinitely divisible, since we are attributing to it the properties of geometrical space. We shall mark on this straight line all the decimal points of division, that is to say, the extremities of decimetres, centimetres, millimetres, tenths of millimetres, &c. We naturally assume that there will be no stopping-place in this *et cetera*, that is to say, that we shall be able to continue *ad infinitum*. We shall now remove a piece of the straight line round each of these infinitely numerous points, points which, to use the suggestive expression of the theory of aggregates, are *dense* over the whole line. What will be left? Surely nothing at all, if we leave the answer to common sense and intuition, seeing that on every piece, however small, of the straight line there is an infinite number of decimal points round every one of which we remove a piece. It is, however, easy to see that we can adjust the size of the pieces we take away in such a way that the portion removed is only a small fraction of the whole line. To show this explicitly, round the points of division corresponding to decimetres we shall remove the length of 1 decimetre *in all*, that is to say, a centimetre for each of the points.[1] We shall in the same way remove a centimetre *in all* round the centimetrical points of division, that is 1/100 of a centimetre or 1/10 of a millimetre for each point. We shall next remove 1 millimetre *in all* round the millimetrical points of division, say, 1/1000 of a millimetre round each point; and so on *ad infinitum*.

[1] In point of fact, there are 11 points of decimetric division, if we count the ends corresponding to 0 and 10 dm.; we must remove only ½ cm. at each end, and 1 cm. symmetrically about the 9 points 0·1, 0·2, . . . 0·9; round 0·4 e.g. we remove the space between 0·395 and 0·405, that is, 1 cm. in all.

It will, of course, sometimes happen that all or a part of the portion to be removed will already have been removed in a previous operation, so that the sum total of the lengths removed will be less than the sum total obtained by adding 1 decimetre, 1 centimetre, 1 millimetre, &c., that is, less than the length which is expressed in metres by the repeating decimal fraction 0·1111..., which is, as we know, equal to 1/9 (a ninth). Now, if the material bar 1 metre long weighs 900 grammes, we shall have removed only 100 grammes, and there will remain 800, although we have removed a small portion round each of the decimal points. I have never yet met anyone who did not agree that this is an impossible result when considered from the point of view of geometrical intuition. It becomes even more so when we try to picture, as we began by doing, the portions taken away and those left as consisting of actual matter. We cannot imagine any *property* of this matter, infinitely cut up in the way described, and constituting, in the language of the theory of aggregates, a perfect aggregate, of measure not zero, and nowhere dense.

What can we conclude from this, except that intuition fails us in the infinitely small as well as in the infinitely great? It is no doubt possible that theoretical physicists may one day utilize these singular aggregates, of which we have just given a simple example, obtained by suppressing an infinite number of intervals, namely the intervals contiguous to the aggregate, as M. Baire has called them. In other words, it is possible that the abstract facts thus formulated may correspond to concrete facts in such a way that they may enable us to explain and to foresee these. But such a transformation of both physics and geometry would require a preliminary revision of our ideas about space, ideas which appear to us to be intuitive.

46. The Sub-atomic Scale.

Without going so far as such a complete revision of our intuitive ideas concerning space, considered on our own scale, it is permissible to ask to what extent it is legitimate to transfer these ideas as they stand, not to the infinitely small, but to the scale which may be called sub-atomic, the scale on which the linear dimensions of atoms would be measured by numbers of the order of a thousand or a million, that is to say, of the same order as the measures of ordinary objects in terms of a millimetre or 1/10 of a millimetre (that is to say, in terms of the smallest practical unit perceptible to us without the help of instruments). On this sub-atomic scale the notion of a solid body on which our geometry is based fails us completely; ought we then to try to construct an entirely different geometry of a discontinuous character? This might be a way of attacking the difficulties raised by the theory of quanta, that most attractive and, one might even say, indispensable theory, which is yet, apparently at least, full of logical contradictions.

Following up a suggestion made by Riemann, to which, however, he did not seem to attach much importance, we might also ask whether the property which space seems to possess, of being in a sense the more certainly Euclidean the smaller the region is which we consider, does not disappear when we reach atomic or sub-atomic dimensions. Very small regions may have, as it were, a granular structure, and the Euclidean properties only hold good as averages. In a similar way, a mirror of polished metal shows on an average the properties of a plane on our own scale, although with a very strong magnifying glass we can perceive considerable irregularities obeying no law whatever.

But before descending, if we may so speak, to these sub-atomic questions, it would seem necessary first to clear up a certain question, at once physical and geometrical, which we may call the postulate of the quadratic form, or the postulate of the ellipsoid. Let us try to state precisely what that postulate is.

47. The Postulate of the Ellipsoid.

If we take three axes of rectangular co-ordinates, Ox, Oy, and Oz, the distance r of a point from the origin of co-ordinates is given by the formula

$$(1)\quad r^2 = x^2 + y^2 + z^2,$$

which is simply the theorem of Pythagoras on the square of the hypothenuse, applied to the diagonal of a rectangular parallelepiped. The equation (1), if r is fixed, whilst x, y, z are variable, will thus represent the equation of the surface of a sphere of radius r, having its centre at the origin; this surface is the locus of one end of a rod of constant length the other end of which remains fixed at O, the rod taking up all possible positions round O. Let us now suppose r, and consequently also x, y, z, to be infinitely small quantities, and let the physical medium on which we are making our measurements undergo some deformation, then we must admit, on grounds of continuity, that infinitely small lengths are transformed linearly. If, therefore, we indicate by r', x', y', z' the new values of r, x, y, z, the equation (1) will become

$$(2)\quad r'^2 = Ax'^2 + A'y'^2 + A''z'^2 + 2By'z' + 2B'z'x' + 2B''x'y',$$

that is to say the equation of an ellipsoid, the quantities $A - 1$, $A' - 1$, $A'' - 1$, B, B$'$, B$''$ being, moreover, supposed to be very small. It is thus that we proceed in the theory of elasticity.

We must, however, point out that in all this we assume that when we are making measurements on the deformed medium, we place it in the absolute Euclidean space which is not deformed. All that we can do in reality is to measure the distances of the actual reference marks by employing as instruments real objects, or, if we prefer it, gauges, the nature of which we need not specify (but which may be a metallic rule or a wave-length). On the other hand, we may regard any continuous system of co-ordinates whatever as defined, when we have actually given the co-ordinates of a very great number of materialized points of reference; the *possibility* of defining such a continuous system is anyhow a primary hypothesis. I confess that I do not see how this hypothesis can be said to imply the postulate of the ellipsoid. It seems to me that the latter cannot be regarded as anything more than the generalization of a certain number of well-known experimental facts. The more or less complicated demonstrations based upon the theory of groups, of which Poincaré has given famous examples and which have recently been considered anew by Weyl, do not get beyond these experimental postulates. The question whether these postulates are rigorously verified, or only approximately so, remains untouched.

48. Geometry and the Quantum Theory.

The theories of electricity and thermodynamics which are actually current show a tendency to substitute the discontinuous for the continuous; the hypothesis of quanta seems to explain phenomena better than hypotheses based on continuity. Does this mean that we should introduce this discontinuity into geometry itself? The question is perhaps somewhat premature, but we can hardly avoid putting it. It is, of course, difficult even to state such a question precisely, but the difficulty, far from discouraging effort, ought to stimulate it. A question well put is often half answered. One of the numerous aspects of this complicated question is, in the domain of mathematics, the problem of the abstract relations existing between the continuous and the discontinuous. May one hope to find in some of these relations a suitable model for representing and predicting certain phenomena? Were we to succeed in this, then the most difficult part of the task would be accomplished, since the first step of bringing the concrete and the abstract into connexion has always constituted the principal difficulty in the application of mathematics to realities. Once this connexion is established at one point, the relatively easy development of abstract theories will rapidly furnish us with new subjects for study and research, and with new analogies.

We shall see presently that these questions of continuity are not less important in topology than in infinitesimal geometry.

49. Maps and the India-rubber Metre.

We have said above that it was not possible to represent a sphere upon a plane correctly, that is to construct a rigorously exact map of a portion of the earth's surface (supposed to be spherical, for the sake of simplicity). If, however, only a small region were in question, we could manage matters in such a way that the errors would be very small. They could even be entirely suppressed, if we were to allow certain conventions to which a mind like that of Poincaré would not be averse. We know indeed many ways of constructing a map of a region of the earth, a map of France, for example, which preserves the shape of very small elements, but the scale of any such map must vary from one region to another. If we were, therefore, to agree to measure distances on the map with a metre which would itself vary according to the region we are considering, we could compensate the difference in scale exactly. We can, for instance, use a metre made of india-rubber tissue instead of one made of tape. This india-rubber metre may be stretched on a mobile metal curtain-rod, one of its ends being fixed to the rod, whilst the other passes freely over a pulley at the other end of the rod and is kept taut by a weight. The weight may consist of a chain falling vertically from the rod and trailing upon the ground. If our map is now placed upon a vertical wall, the length of the chain will vary (a longer or shorter portion trailing on the ground), and consequently also the weight stretching the metre, according as we are using the metre to measure an upper or a lower part of the map. It is not difficult to conceive that an exact compensation might thus be produced between this variation in the length of the metre and the variation in the scale of the map, so that if we consider the length of the metre as constant by definition, then the scale of the map will also be constant.

This compensation, however, is possible only for a map reproducing no more than a part of the earth's surface. If we were to represent the whole earth, using, for example, Mercator's projection, our india-rubber metre would have to be lengthened *indefinitely* in the neighbourhood of the pole. Leaving out the question of the limits of elasticity, we could not pretend not to notice that its length had altered. Anyhow, for the pole itself, this length would have to be really infinite, which is, of course, impossible.

50. Discontinuity Inevitable in a Plane Map of a Sphere.

One might be inclined to think that this difficulty is due to the particular choice of Mercator's projection, and that we could obviate it by the choice of another projection. But it is easy to show that such is not the case, and that it is impossible to construct a representation of the sphere on the plane *with no singular points and without discontinuity*. Let us try to explain in ordinary language the reason for this very important result. Let us start from some point of the earth, from Paris, for instance, in an aeroplane or in an airship so as to meet with no obstacles. We can easily mark on the ground the totality of the points which we might reach, supposing we leave Paris in all possible directions, after an hour, two hours, twenty-four hours, two days, and so on. We can agree to say that the totality of the points we reach at the end of an hour constitutes a circle of which Paris is the centre and the radius of which is equal to an hour. It is clear that the circle, the radius of which is equal to two hours, is larger than the circle the radius of which is equal to one hour, surrounding it entirely. If a map is to be of any use, the least we can demand of it is that these circles be each represented by a closed curve surrounding Paris, the curve corresponding to two hours surrounding completely the one which corresponds to one hour. We must thus have successive curves round Paris, curves which we can number after the number of hours as 1, 2, 3, . . . 24, 48, &c. These curves, in any map drawn on a plane, ought to increase steadily in size, the greater always surrounding the less.

But, supposing the radius of action of our airship to be large enough, it will follow that if it travels a distance of 200 kilometres per hour it will have travelled 20,000 kilometres in 100 hours, and so have reached the antipodes of Paris, whatever direction it may have followed.

The curve numbered 100 on our map (fig. 7), a curve which should surround all the others, will thus correspond on the sphere to one point that is antipodal to Paris. Here we have one of the singularities or discontinuities which we have already mentioned as bound to occur. We see at the same time that in this particular case the discontinuity is unique, in the sense that it is relative to a single point upon the earth's surface, the point antipodal to Paris.

The discontinuity, however, will also affect every curve passing through this point and will modify neighbouring curves profoundly.

As might be anticipated, we can avoid the discontinuity at this point by transferring it elsewhere. If it takes us 100 hours to reach the antipodes, we shall arrive in

50 hours at the various points of a great circle of the sphere of which Paris is the pole. We can represent this circle both by 50 and by 50′ (figs. 8 and 9), and then trace the

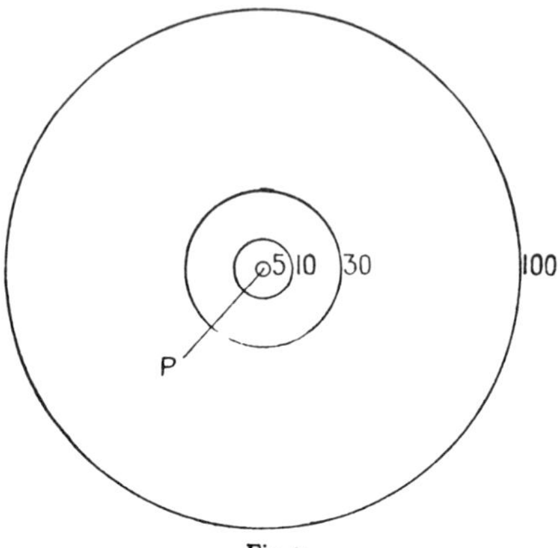

Fig. 7

succeeding circles in the interior of circle 50′ instead of outside it. We thus trace the circles 55, &c., 90, 95, the circle 100 being reduced to the point P′, antipodal to P. We see that the discontinuity will here consist in the points

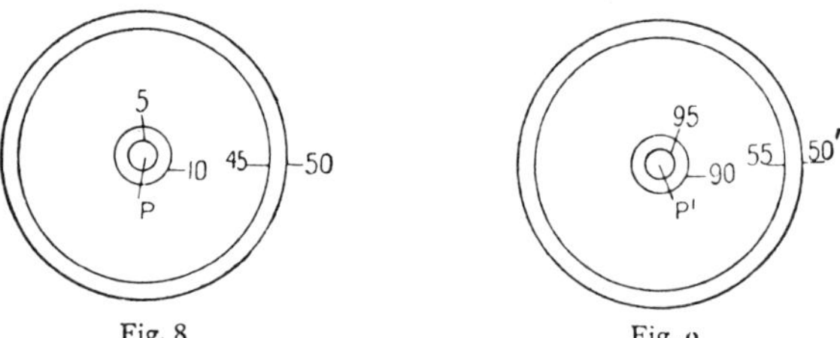

Fig. 8 Fig. 9

of the circles 50 and 50′ being separated instead of coinciding in pairs. We may, if we like, cut one of these circles out and paste it against the other. We shall thus have a continuous representation of the sphere upon a double plane sheet, not on a single plane.

51. A Sphere has no Boundary.

We can prove in an even shorter and more synthetic manner the impossibility of a continuous correspondence between the sphere and the plane. If we are dealing with the whole plane the impossibility is evident, for the plane has infinite distances, whilst on the sphere all distances are finite; if only a portion of the plane is in question, this portion will be bounded by an edge or border. Now the sphere is a finite but yet unbounded surface having no border; and only *one* point of the sphere will have to correspond to the edge of the plane. This is one of the forms of discontinuity which we have already encountered.

52. Topology of the Anchor-ring.

This topological point of view, from which we examine the relations of situation between the various points of a surface, taking account of continuity only and not troubling about exact measurements of distances, is of great importance in the formation of our notion of space. Inhabitants of the earth who possessed neither geometrical nor astronomical knowledge, might by patient exploration find out that our globe is finite and has no borders. We must, of course, assume that their explorations would not be stopped by obstacles, whether polar ice, the deserts, or the oceans. They would also find that if they were to construct—or to imagine constructed—a continuous wall in the shape of a closed circuit it would be impossible to pass from one side of it to the other without crossing it. This property, which in the actual case seems a self-evident truism, would not hold good if the earth had the shape of a tore, or anchor-ring. If we take, for instance, a loaf of bread in the shape of a wreath, and placing a knife inside the ring make a cut towards the outside, we do not detach one piece from another. The loaf when cut through will still form a single piece, and it would require another similar cut to divide it into two parts. The cut made with the knife in the crust of the bread is a closed curve, but it is not a boundary, for if very small beings living upon the bread were to erect a wall, an ant could pass from one side of the wall to the other without crossing it. It would only have to go round the ring, keeping on a horizontal plane, supposing the loaf to be placed upon a table and the cut to be vertical.

53. Local Knowledge cannot give Knowledge of the Universe.

Topological ideas are intuitive when it is a question of the plane, of the sphere, and of the tore or anchor-ring, that is to say of two-dimensional domains which we locate in our three-dimensional space. As we are not endowed with any direct geometrical intuition of four-dimensional space, we have not the same ready capacity for imagining the infinite variety of three-dimensional domains in space of four dimensions, domains of which analytical geometry furnishes us with algebraical representations. Thus intuition leads us to think of the space in which we live as possessing with its three dimensions the same properties as the plane of two dimensions. But if we reflect a little we shall find that we have no more solid reason for believing that space is so constructed than the ancients had for believing the earth to be flat. Had the earth been flat, it would either have been unlimited or bounded by an edge; and the same is true of our own space if it be Euclidean. But the idea of an edge is repugnant to ordinary common sense, since the question always suggests itself: What is beyond the edge? Euclidean space is, therefore, generally considered as unbounded, in spite of the objections of the philosophers to the idea of the actual infinite. It is, however, less easy to overcome these objections, when it is a question of the real universe, than it is when we are dealing with abstract space. Mathematicians have noticed for a long time that there is no contradiction in the conception of a finite space with no edge or border. We can imagine such a space under the analytical form of a three-dimensional sphere in four-dimensional space. The radius of the sphere would then be what we call the radius of curvature, the reciprocal of which is the curvature itself. In such a space the shortest lines are analogous to the great circles of the terrestrial sphere; they are closed, their length being equal to the product of the radius by 2π. It is clear that if the radius of our universe be equal to some thousands of millions of light-years, then we possess no experimental means for finding out whether the lines we are calling straight are not in reality closed circles of very great radius. To make this hypothesis is not more paradoxical than it would have been for a primitive man to suppose that by constantly walking straight on in a due easterly direction he would finally return to the point he had started from, arriving at it from the west.

It is, of course, understood that when we say that this hypothesis is not absurd we have no intention whatever of asserting that it is true or even plausible. In reality, we have no more reason to believe it to be true than we have for not believing it. Before the immensity of space we are in the position of a naked man upon a rock in mid-ocean, under a cloud-covered sky, and unable to observe the stars. Supposing that attentive observation of the surface of the ocean suggested to such a man the idea that the portion of the ocean in his immediate neighbourhood is rather convex, what right would he have to conclude that he is on a sphere? At that rate, a mountaineer who had never left the neighbourhood of a lake surrounded on all sides by high mountains might conclude that the earth was a concave basin.

54. The Plane Topological Representation of a Sphere.

With the preceding reservations, there is no harm in trying to form an idea of what a non-Euclidean space would be like, that is to say, in trying to express in a geometrical form, appealing to our senses, the analytical conception of such a space provided so easily by algebraical formulæ. In order to make our procedure more intuitive, we shall apply it first to the passage from the plane to two-dimensional surfaces situated in the space to which we are accustomed. In other words, we shall consider how we might convey the idea of a sphere or of a tore to a being who was confined to two dimensions and could only conceive the Euclidean plane.

We have already said that, topologically, the sphere is equivalent to a double-plane sheet. If we cut out two equal circles in thin india-rubber and paste them together along their circumferences, we shall obtain a surface composed of two plane sheets. Supposing that we inflate this surface as we would inflate the inner tube of a bicycle tyre, then its shape, when the pressure is sufficient, will become approximately spherical.[1] We can imagine a geometrical sphere which has its centre O in the interior of this quasi-sphere of india-rubber, and make each point A of the geometrical sphere correspond to the point A′ of the quasi-sphere where it is pierced by the radius OA. This point A′ is either a point A_1 of one of the plane sheets, or a point A_2 of the other sheet, or a point B of the common circumference, a point which we may call B_1 on the one sheet and B_2 on the other. We may thus say that, topologically, the geometrical sphere is equivalent to the combination of two circular sheets, provided we consider two corresponding points B_1 and B_2 of the two circumferences as *one* point; these two corresponding points B_1 and B_2 coinciding when

[1] On account of the fact that the sphere is the surface of minimum area enclosing a given volume.

we paste the circles against each other. Suppose next that our india-rubber sheet has a right side and a wrong side. Naturally, we will wish the surface of our sphere to show the right side everywhere; for this it will be necessary to have the wrong sides of the two sheets facing each other. Having, therefore, cut out the two circles from one plane sheet, we shall have to turn one of them over, or, what amounts to the same thing, substitute for it its image as shown in a mirror. When the two sheets are placed against each other, corresponding points, such as B_1 and B_2, or C_1 and C_2, will coincide, and the arcs B_1C_1 and B_2C_2 will not only be equal but will have the same sense; that is to say, if we think of each of these circles as the dial of a clock, the minute hand would have to advance five minutes in order to go either from B_1 to C_1, or from B_2 to C_2, assuming we are looking at the two sheets so that B_1 and C_1 are on top. If, however, we turn B_2C_2 back again so as to show the right side, the minute hand would have to go five minutes backwards in order to go from B_2 to C_2. If we look at the sheets at the moment when we have cut out the circles, when their right sides are visible on the same side of the plane, points which correspond will be symmetrically placed with respect to a certain axis xy. We might, therefore, carry out the construction by folding the sheet about the axis xy, adjusting the two halves to fit each other, and then cutting it out. If we consider the three points A_1, B_1, C_1 and the corresponding points A_2, B_2, C_2, the elements of the triangles thus formed will be respectively equal, nevertheless the triangles cannot be made to coincide unless the plane is turned over. If we were merely to make the plane slide along itself, we could no more fit the one triangle on the other than we could fit on one another the outlines traced with a pencil round a right and a left hand placed palms downward on a sheet of paper.

If we wish the combination of two plane circles to furnish us with a topologically exact image of a sphere, we must make an effort of abstraction and consider the points B_1 and B_2, or C_1 and C_2 as identical and coincident. The two curves, for instance, B_1MC_1 and C_2NB_2 must be considered as constituting *one* closed curve which may be called BMCNB, the letter B indicating the *unique* point B_1B_2, and the letter C the unique point C_1C_2.

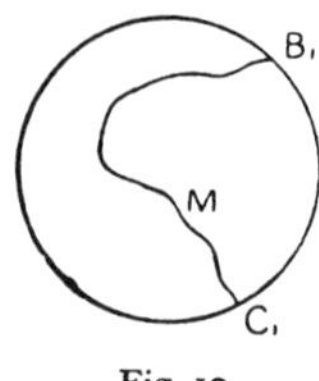

Fig. 10

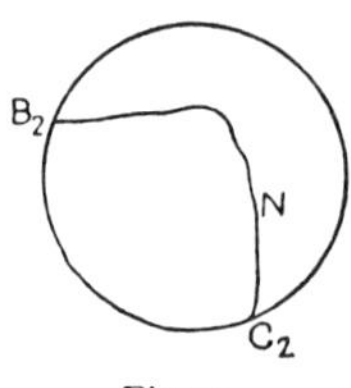

Fig. 11

The above method of picturing the surface of a sphere is the one used in the familiar type of map of the world which consists of two circles placed side by side to represent the two hemispheres.

55. Topological Representation of a Hypersphere.

Now just as we have obtained by means of two circles a topological image of a sphere, a two-dimensional surface situated in three-dimensional space, so also we can obtain, by means of two spheres, the topological image of a three-dimensional *hypersphere*, which, from the Euclidean point of view, would require a four-dimensional space to serve as its support. The intuitive geometrical image fails us here, although we could give an exact algebraical representation of the arguments to which analogy leads us. It will be sufficient for the reader to know that this algebraical image fully confirms the generalization by analogy to which intuition naturally leads us. Let us, therefore, consider two spheres each of which is supposed to be the reflection of the other in a mirror, and let us regard corresponding points (reflections of each other) of the surfaces of these spheres as being identical and coincident. With this assumption, the combination of the two spheres will give an exact topological image of the three-dimensional surface of the hypersphere. It would be easy to complete this topological image by means of calculations and to make it metrical, that is to say to give precise definitions of distances and angles, but we shall not wait to do this, as it would add nothing to our geometrical description. On the other hand, it will not be superfluous to dwell a little upon the correspondence existing between the points of the surface of the two spheres, one of which is the reflection of the other in a mirror. We know that two geometrical figures in space, symmetrical with reference to a plane, are not identical; the reflection of a right hand in a mirror is a left hand. Two spherical triangles ABC, A'B'C' which are situated respectively on the two spheres and have this kind of symmetry cannot, therefore, be superimposed, not even if they are infinitely small with reference to the radii of the spheres. How should we, therefore, picture space in a region the image of which is near the surface of the two spheres? Let us imagine a body to be changing its position within the first sphere and to reach a point within the triangle ABC. If it continues its path in the interior of the second sphere, it will be passing through the triangle ABC from inside to outside, while in the case of the triangle A'B'C', it will be moving from outside to inside. This means that, assuming the surfaces of our two spheres to be materialized (like a balloon of gold-beater's skin), the outside of the one will correspond to the inside of the other, and conversely. We may, for instance, assume the one sphere to be painted black inside and white outside, whilst the second is black outside and white inside. Under these conditions, if we were to cut out a triangle ABC on the first sphere, we could apply it to the symmetrical triangle A'B'C' on the second sphere, provided we first turned it over, that is to say brought the black-painted inner side to the outside.

56. A Finite but Unbounded Universe.

If then space were hyperspherical and not Euclidean, this is what might happen. If, under the guidance of a being possessing means of action in space and time much superior to ours, we journeyed in a straight line in any direction, then, after travelling for several millions or thousands of millions of years with the velocity of light, we would reach the points of a certain "sphere". Continuing our way for an equal time without altering our direction, we would reach *a second centre* of this sphere, the antipodes of our starting-point. If we were to continue again we would end by arriving at the point we had started from. The universe would then be finite and without boundary. There is nothing logically absurd in such an hypothesis, although there is also nothing in its favour. The hypothesis, however, is not more unreasonable than the Euclidean hypothesis.

57. The Ring and a Plane Network of Rectangles.

Another image which has frequently been suggested for the universe, and which has given rise to certain misunderstandings, is that which we obtained by generalizing the tore or anchor-ring. Let us first point out how we can construct a tore by means of a rectangle. Cut out of a sheet of india-rubber a long rectangle, and paste one of the two longer sides against the other. We shall thus obtain a long cylindrical tube terminating at each end in a circle formed by a small side of the rectangle. If we now paste these two circles against each other, we get a ring analogous to the hollow tyre of a bicycle. If now an ant were to walk along on the surface of this ring, it would be able to come back to its starting-point in many ways, two of which are specially noteworthy. It might describe a small circle equal to one of the two small circles which were pasted together, thus travelling a distance equal to one of the small sides of the rectangle. The ant could also describe a large circle practically equal to one of the long sides of the rectangle,[1] the circle, for instance, which rests on the

[1] These large circles are not all equal, the india-rubber being necessarily distorted, but there is nothing to prevent us from regarding them (by convention) as equal.

ground when we are inflating the tyre, keeping it in a vertical plane.

We may thus consider the tore or ring to be topologically equivalent to a plane rectangle, provided the opposite sides of the rectangle are considered as coinciding two and two. In the rectangle ABCD (fig. 12), the point M coincides with the point M′, and P coincides with P′; a continuous curve ending at M must be considered as continued by a curve starting from M′.

We can avoid this discontinuity, which makes us jump from M to M′, or from P to P′, by the following geometrical device. Let us cover the plane with a network of equal rectangles, and let us agree that two points are to be considered *identical* if they are similarly placed within two of these rectangles, for instance, two points such as R and R′, in the neighbourhood of a lower left-hand corner and placed at an equal distance from this corner. Under these conditions, the curve RSTUV shown in full lines will be equivalent to the curve parts of which are shown dotted inside the first rectangle. The two curves have RS in common, S′T′ corresponds to ST, T″U″ to TU, and U″V′ to UV. From a topological point of view it is exactly the same whether we consider *one* rectangle, agreeing upon the equivalence, accompanied by continuity, between points such as SS′, T′T″, U′U″; or consider the indefinite network of rectangles, agreeing upon the equivalence of corresponding points. If we lived on a tore and, not knowing any other but plane geometry, tried to get an idea of these phenomena, we could, by classifying the different ways of returning to our starting-point, explain the facts observed by means of the following peculiar hypothesis: There

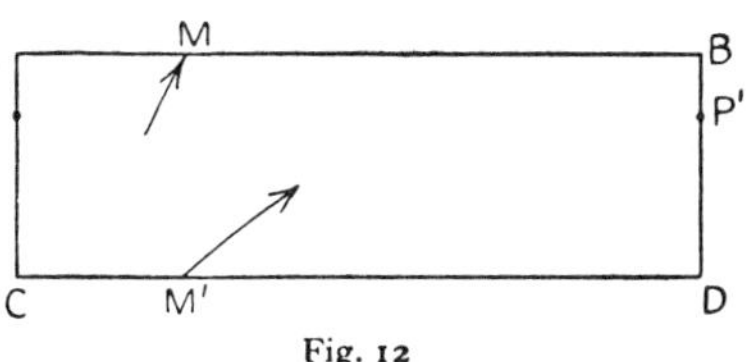

Fig. 12

exists an infinity of identical rectangles having the property that all that occurs within one of them is repeated at the same instant within all the rest. If we were to walk straight on we would never come back to the point we started from, but would arrive at one of its homologous points.

One might ask whether this singular hypothesis, impossible to demonstrate, could possibly be accepted, or whether it should not be considered as being absolutely identical with the more simple hypothesis that we live on a tore.

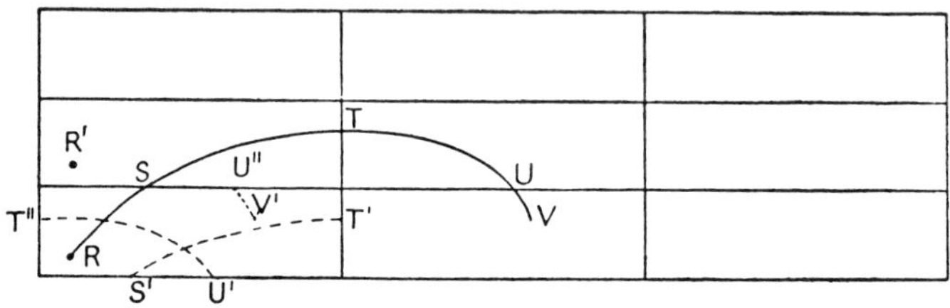

Fig. 13

I think, however, that we should be leaving here the domain of science and entering upon that of metaphysics. Let us say, however, that it would not be a sufficient reply to point out that the development of science may, perhaps, enable us some day to perceive simultaneously two or more identical points and thus to prove that they are actually distinct. Theoretically, there is nothing against the possibility of our perceiving simultaneously, by means of a play of mirrors, two reflections of the Eiffel tower, one of these reflections being seen directly, whilst the rays coming from the other have gone round the earth. This would be no proof of there being two earths, and the same point of view could be taken if the earth were a tore and not a sphere.

58. The Hypertore and a Periodic Image of the Universe.

The details just given allow us to abridge the extension to space of this idea of a tore obtained by means of a rectangle. We should have to take a rectangular solid and consider as equivalent the points of any two opposite faces which lie on a line perpendicular to these faces. By performing the ideal operation in four-dimensional space of pasting the equivalent points together, we obtain a sort of *hypertore*, the topological study of which appears, at first sight, to be sufficiently complicated. It can, however, be simplified by reducing it to its origin, that is to say by noticing that it is identically the same as a rectangular solid, the points of whose surface are coupled two and two; or, if we prefer, identically the same as a whole network of equal rectangular solids, such that the internal points of two of them are taken to correspond, if they come into coincidence by the *translation* which brings about the coincidence of the two rectangular solids themselves.

We thus obtain an image of what may be called a triply periodic universe; that is to say, one in which there exist three directions such that if we travel in a straight line in one of these directions over a definite distance (which is not necessarily the same for each of the three directions), we reach a spot in the universe identical with the one we started from. It is, of course, understood that this identity implies that a new displacement equal to the first will again give us an identical place and so on *ad infinitum*. In this form the hypothesis seems odd enough, perhaps even absurd. It might, perhaps, seem rather more reasonable if we were to admit that in reality we reached not the identical place but an exactly *similar* one. This means that the periodicity is of the kind which would be found by an ant walking over a ring, and not that which would be found by the self-same ant walking along a vast wall, the design of the wall-paper on which repeated itself indefinitely.

CHAPTER V

The Propagation of Light

59. Fresnel's Theory and the Sinusoid.

We have found it impossible to define space and time physically, that is to say, to measure them, without having recourse to the properties of light. It is, therefore, essential to summarize, as rapidly as possible, these properties.

It was Fresnel who, at the beginning of the nineteenth century, gave us the first coherent theory of optical phenomena, a theory which accounted for all the phenomena then known and made it possible to predict new ones, which *a priori* appeared highly improbable to Fresnel's contemporaries, but which have since been verified by experiment. In the course of the nineteenth century, Fresnel's theory has been several times developed and modified. New phenomena have been discovered necessitating modifications, sometimes slight, sometimes profound. One may, however, say that the theory still holds good; it even continues to be one of the most perfect models of a physical theory, provided at least we regard only its mathematical skeleton, without troubling to give the equations any interpretation other than that which results from their experimental confirmation.

Thus reduced to its essentials, Fresnel's theory may be summed up as follows. The properties of a ray, or rather of a wave, of light are represented exactly by the properties of a simple trigonometrical function, sine or cosine.

The argument of this trigonometrical function, that is

to say, the variable it depends upon, is a linear function of time and distance. Translated into ordinary language, this means that the properties of light are comparable with those of the waves produced by the fall of a stone in the middle of a calm pond. We can study these waves either at a fixed point, by observing a piece of cork rising and falling as the wave passes over it, or by concentrating our attention upon one of the waves which appears to be moving and growing larger at a definite speed. Both movements can be exactly represented by a *sinusoid*, that is to say by a curve such as ABCDEFG . . ., which extends indefinitely without change of form in both directions. The length

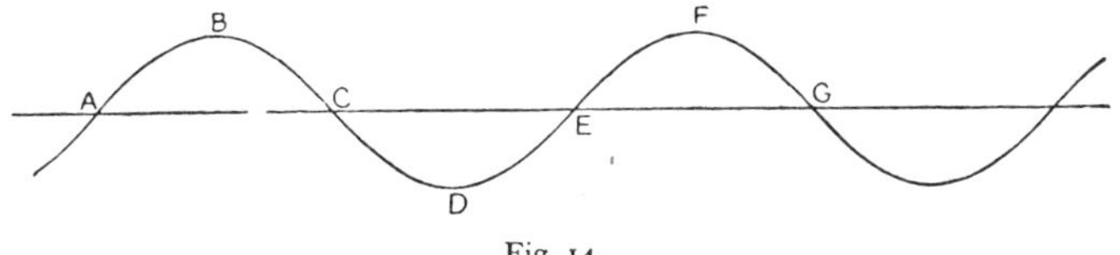

Fig. 14

AE is the *wave-length*, and the time required for the undulatory movement to traverse this distance is the *period*. The wave-lengths and the periods are *physical realities* and may be determined by experiment in various ways; the sinusoid is a mathematical representation of these physical realities. For our purpose, this is all we need to know. There is no need for us to ask whether a fluid really exists, with properties which cease to be mysterious only to become contradictory, a fluid which may be called the luminiferous ether, and which actually vibrates, either transversally or longitudinally. Nor need we ask whether it would not be preferable to interpret these vibrations as being those of an electro-magnetic vector, or of an electron, or the periods of revolution of extremely small corpuscles in the interior of an atom turning round a central nucleus. Whatever the answers may be to these questions which have rightly engrossed physicists for a century, they do not alter the methods by which wave-lengths and periods can be measured.

60. Wave-length and Difference of Phase.

Let us consider two sinusoids referred to the same axis *OX*. We assume the two sinusoids to be equal, but one of them, which we shall represent in dotted lines, is retarded with respect to the other by a fraction of a period. If we suppose the retardation, or *difference of phase*, as it is called, to be exactly half a period, it will be clear from the diagram that at every point M of *OX* the ordinate MP′ of one of the sinusoids will be equal and of contrary sign (or, more

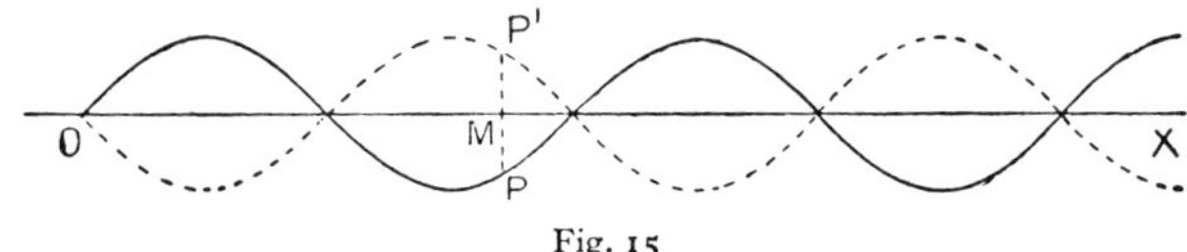

Fig. 15

briefly, opposite to) to the ordinate MP of the other sinusoid. If we now admit with Fresnel that an optical phenomenon observed under certain conditions is measured by the *sum* of these ordinates, this phenomenon will be null; that is to say, light added to light under certain conditions will produce darkness. This is one of the results predicted by Fresnel, and one which appeared very paradoxical to common sense. It has, nevertheless, been fully proved by experiment, and the use of *interference* has now become one of the surest and most effective means of investigation in physics.

This is not the place to develop, even summarily, this important branch of science, which is called physical optics, and which was revolutionized by Fresnel's epoch-making work. Let us merely state that the means by which differences of phase are obtained, the retardations produced by interference, enable us to measure the wave-lengths of light vibrations of every kind. This measurement may be effected by two quite distinct methods, one of these methods enabling us to compare one wave-length with another, that is to say, to measure all wave-lengths in terms of one which is arbitrarily taken as the unit. These methods of comparing wave-lengths among themselves were long ago brought to a high degree of precision. On the other hand, it is only within the last few decades that we have been able to measure wave-lengths with great precision in *absolute measure*, that is to say, to compare them with a standard measure of length, namely, the international metre of the Pavillon de Breteuil.

61. Measurement of Wave-lengths in Metric Units.

Without entering into the details of interference methods, we may merely mention the fact that it is by counting the *fringes*, that is to say, the alternately light and dark bands due to interference, that we are able to measure an assigned length by means of a wave-length chosen as the unit. As soon as we know how to count a hundred fringes, there is no theoretical difficulty in counting a thousand, a hundred thousand, or even several millions of fringes. In practice, however, apart from the fact that it would require a considerable time to count one by one such big numbers, it would be almost impossible to make sure that no error had been made in the counting. This difficulty has been got rid of by Michelson by means of a very ingenious device, consisting in the simultaneous employment of two different wave-lengths. The appearances observed enable us to determine at what moment the length which is being measured is an integral multiple of both these wave-lengths. On the other hand, optical methods of the same nature have enabled us to determine with very great precision the ratio of these two wave-lengths, and simple arithmetical calculation makes it possible to deduce the exact number of wave-lengths contained in the length to be measured from an approximate value of this number, which may be supposed to be known, or at any rate can be found by means of ordinary micrometrical operations. By combining these various methods and dividing up the process into a sufficient number of intermediate steps, it has become possible to estimate the number of wave-lengths contained in the international metre of Breteuil. The precision attained is of the same order as the precision with which this metre itself may be considered as defined. The wave-lengths are thus known in absolute measure. They are usually estimated by means of a metrical unit to which the term *ångström* has been given, equal to 1/10,000 of 1/1000 of a millimetre, that is to say, the ten-thousand-millionth part of a metre, more briefly described as 10^{-10} of a metre, or one *tenth-metre*. The experiments of Benoit, Fabry, and Pérot on the measurement of the metre in wave-lengths by the method of Michelson show that the red radiation emitted by cadmium vapour contained in a vacuum tube has a wave-length, in dry air at 15° C. and under normal atmospheric pressure, of a little over 6438, or, more exactly, 6438·4696 ångströms. The wave-length of the same ray *in vacuo* is a little longer, being 6440·249 ångströms. The *frequency*[1] is generally defined as the number of wave-lengths *in vacuo* contained in a centimetre. As a centimetre corresponds to 100 millions of ångströms, the frequency is found by dividing 100 millions by the wave-length *in vacuo* expressed in ångströms. For the red ray of cadmium it is equal to the quotient of 100 millions divided by 6440·249, that is to say, to 15527·34. As for the number of vibrations per second, we obtain it by multiplying the frequency by the velocity of light *in vacuo*, expressed in centimetres, that is to say, by about 30,000 millions. For the red ray of cadmium, this number of vibrations per second is equal to about 466 million millions. We could define a

[1] In Britain, the number of wave-lengths per centimetre *in vacuo* is usually called the *wave number*, the term *frequency* being used for the number of vibrations per second.

unit of time as being exactly equal to the duration of this vibration of a well-determined ray. We see that the unit thus specified is about the double of the one-millionth part of the one-thousand-millionth part of a second.

It is, however, important to notice that the most uncertain number in these evaluations is that which expresses the velocity of light, the values of the wave-lengths, and consequently of the frequencies, being known with much greater precision than this velocity. It is, indeed, very difficult to measure the velocity of light in absolute measure. Relative measurements, as we shall see, can be made much more easily, by employing differential methods to compare with each other either the velocities in different media or the velocities in different directions in the same medium.

62. Measurement of the Velocity of Light.

The methods employed for measuring the velocity of light may be divided into two great categories: purely optical methods, and electromagnetic methods, the latter assuming numerical identity between the velocity of light and the ratio v of the electromagnetic unit of quantity of electricity to the electrostatic unit. The optical methods are again subdivided into two great categories, astronomical and terrestrial methods. The astronomical methods assume the dimensions of the solar system to be known and would, conversely, enable us to calculate these dimensions in terms of the velocity of light. We have already spoken of the observations made by Roemer (1675) on the eclipses of the satellites of Jupiter, and we shall soon have occasion to say a few words (§ 67) with regard to the *aberration* of the fixed stars, as observed by Bradley in 1727. The terrestrial methods consist in measuring directly the time required for the light to traverse, to and fro, a certain distance measured directly upon the earth's surface. These latter methods reduce to two, both devised by French physicists, the toothed wheel method (Fizeau, 1849), taken up again by Cornu (1871–4) and by Perrotin (1904), and the revolving mirror method (Foucault, 1849–62), taken up again by Michelson and Newcomb (1885), and by Michelson alone (1902). Here are, after the table given by G. Lippmann in the *Recueil des constantes physiques*, published by the French Physical Society, the numbers obtained by these different observers and also by other observers who have employed electromagnetic methods. The numbers are given in *thousands of kilometres per second*, the last figure being increased by one when the first figure omitted reaches or exceeds 5; numbers which have been treated in this way are indicated by an asterisk.

Optical Methods.		Electrical Methods.	
Roemer	299*	Ayrton-Perry ..	298
Bradley	298	Stoletow	299
Fizeau	313	J. J. Thomson ..	296
Cornu	300	J. Clemenčič ..	302*
Perrotin	300*	Himstedt ..	301*
Foucault	298	H. Abraham ..	299
Michelson-Newcomb	300	Hurmuzescu ..	300
Michelson ..	300	Rosa-Dorsey ..	300

From the combined figures it results that the value of 300,000 kilometres per second may be adopted with sufficient safety, but that the actual accuracy of this measurement is, at the utmost, of the order of 1/1000 of the measured quantity. More recent measurements give 299·9* (Michelson) and 299·7 (Rosa and Dorsey). To this degree of precision there is no need to distinguish between the velocity in air and the velocity in empty space. The ratio of these two velocities is equal to the index of refraction of air, which is 1·0002929, or, say, 1·0003; that is to say, the velocity in air is smaller than the velocity in empty space by about 3/10,000 of the latter, i.e. by about 90 kilometres per second. We may emphasize the fact that the ratio of the velocities in air and in empty space is known with greater precision than the velocities themselves.

63. Measurement of Very Short Intervals of Time.

Thanks to interference methods, the difference of the times required for light to traverse two different distances can also be estimated with extreme accuracy. The degree of accuracy may correspond to a fraction of a wave-length; but a wave-length itself corresponds to the period of a vibration, that is to say to the fraction of a second obtained by dividing the wave-length by the velocity of light. This quotient, as we have seen, is of the order of a millionth part of the thousand-millionth part of a second, or, if we prefer, to 10^{-15} of a second. If now we remember that a duration of 10^{15} seconds is but little less than a million centuries, we see that there are as many vibrations of light in a second as there are seconds in the longest geological periods. If there were beings for whom time passed slowly enough to allow them to distinguish from each other the moments of two consecutive vibrations, a second to them would be what the geological periods are to us. It is remarkable enough that both the longest and the shortest durations which we are able to estimate should stand in almost the same ratio to the duration of a second, a normal unit of duration on the human scale (1/10 of a second represents the shortest interval which we are able to perceive, or even in any real sense imagine).[1] If we wished to indicate precisely how the very small lengths and durations of which we have just spoken can be measured, we should have to enter into developments far exceeding the limits we must set ourselves in a work which is not a treatise on physical optics. It is sufficient for us to say that just as the measurement of the distances of the stars and the chronology of celestial phenomena imply an immense number of observations, corrections, and calculations, and, in fact, are based upon and bound up with numerous mathematical and physical theories, so also the

[1] See the communication of M. Charles Richet to the Academy of Sciences, Dec., 1921.

measurements of very small lengths and durations cannot be separated from an extremely complex combination of experiments and theories.

64. X-rays and Crystal Structure.

In the course of the last few years the interference method has been considerably, and unexpectedly, extended to the study of the action of crystals on X-rays (Röntgen rays). This action, discovered by the German physicist von Laue, has been studied particularly in England by Sir William Bragg and his son, W. L. Bragg, and in France by Maurice de Broglie. Rays of light give rise to interference phenomena when they are diffracted by a grating, consisting of a perfectly plane surface, usually of glass, very closely ruled with a series of equidistant parallel lines. The lines must be the closer the shorter the wave-length to be measured. Granting that the X-rays are radiations whose wave-length is about 1/1000 of the wave-length of the luminous radiations perceptible by the human eye, then, in order to diffract them, we should require a grating in which the distance of the ruled lines would be of the order of molecular distances. It is physically impossible to construct such a grating, but, thanks to the regular disposition of their molecules, crystals in themselves realize a geometrical arrangement sufficiently like a grating for the X-rays traversing them, or reflected at one of their plane faces, to give rise, under certain conditions, to interference phenomena. The deeper study of these phenomena has enabled us in certain cases to arrive at definite conclusions with regard to the geometrical arrangement of the different atoms of which the molecule of the crystal is built up. No doubt, such a static representation cannot pretend to be an exact image of reality, but it supplies us, at least, with a simple geometrical scheme which has numerous properties in common with a real crystal.

65. Michelson and Morley's Experiment.

We have just said that the determination of the velocity of light has been attempted by many methods, both optical and electromagnetic. Without entering upon details of these methods or discussing the principles and postulates on which they are based, we can see that the determination must necessarily involve a complicated experiment, the interpretation of which assumes the acceptance of certain physical theories. Moreover, in the terrestrial optical methods we measure the time required by the light to traverse twice, that is backwards and forwards, the distance separating two points fixed relative to the earth. The question naturally suggested itself whether the measurement of the velocity of light would enable us to demonstrate a motion of the earth relative to the ether, the supposed fixed medium in which the phenomena of light take place. It is clear, indeed, that if optical phenomena take place in a motionless ether, relative to which we ourselves are in motion, our speed will have to be added to or deducted from the velocity of light relative to the ether, according as we are moving in one sense or in the opposite. In the following chapter we shall return to the elementary kinematical arguments leading us up to this conclusion, and we shall endeavour to appraise the value of these arguments. For the present, however, it is sufficient to call to mind the famous experiment by which Michelson and Morley showed that no drift relative to the ether could be proved by utilizing the rotation of the earth round the sun, thanks to which, at intervals of six months, our globe possesses velocities of 30 kilometres per second in opposite directions, relative to the sun.

66. Michelson and Morley's Experiment.

The principal difficulty in demonstrating the motion of the earth relative to the ether, the hypothetical support of the phenomena of light, consists, as we have said, in the fact that terrestrial optical methods are not capable of making a precise measurement of the time required for light to travel from one point to another, but only of the duration of a to-and-fro journey. In any case, the precision of which this measurement is susceptible would be quite inadequate. Thanks, however, to the effects of interference, we can measure with very great precision indeed the difference of the times required for the light to make two journeys out and in, in the case when the two distances are situated at right angles to each other. If the times are equal, we must conclude that the two distances are also equal, if we ignore the possible influence exercised by the motion of the earth relative to the ether. This equality once obtained, it will be sufficient to turn the two distances round into a different position relative to the earth. If the equality remains, we shall have to conclude that the earth is motionless relative to the ether. If the experiment is repeated six months later and gives the same result, the hypothesis, according to which the velocity of light is supposed to be influenced by the velocity of the earth relative to the ether, would lead us to the geocentric hypothesis, which is contradicted by the whole system of celestial mechanics. We must, therefore, conclude that the velocity of light, estimated in a system in motion by means of measurements made within this system, has a value which remains the same in all directions, and which is consequently independent of the motion of the system. Such is the result of the experiment of Michelson and Morley.

In this rapid exposition we have omitted the simple kinematic argument with regard to a journey to and fro, the duration of which is not the same in all directions when the body is carried along by some current or other, in addition to having a definite velocity of its own. Supposing a swimmer whose own proper speed is 40 metres per minute to be swimming in a current the velocity of which is 30 metres per minute, his resultant speed relative to the banks will be 70 metres per minute when he is going with the stream, but only 10 metres per minute against the stream. In order to swim a distance of 70 metres each way up and down stream, he will require in all 8 minutes, that is to say 7 minutes to swim against the current and 1 minute to swim with it.

If, on the other hand, the swimmer proposes to cross the stream at right angles to its direction, and does not wish to be carried down stream, he will have to aim at a point on the opposite bank up stream from his starting-point. In reality he will describe in the course of a minute one side of a right-angled triangle, the hypothenuse of which is 40 metres (corresponding to the swimmer's own speed) and the other side of which is 30 metres (corresponding to the speed of the current); the square of the length of the side at right angles to the stream is equal to the difference between the squares of 40 and 30, that is to say 700. This side is therefore equal to $10\sqrt{7}$ metres, so that the apparent speed of the swimmer is $10\sqrt{7}$ metres per second. In order to swim twice across stream, say a distance of 140 metres, he will require a number of minutes equal to the quotient of 140 by $10\sqrt{7}$, that is $2\sqrt{7}$ or 5·29, a much shorter time than the 8 minutes required to cover the same distance up and down stream.

In the Michelson and Morley experiment the case where the effect would be weakest is that in which the sun is supposed to be motionless relative to the ether. The velocity of the earth relative to the ether would then be 30 kilometres per second, that is to say about 1/10,000 of the velocity of light. With regard to the times required for the to-and-fro journeys in the two perpendicular directions, one in the direction of the earth's motion and the other perpendicular to it, an easy calculation brings out a difference of the order of one hundred-millionth part. The precision of the optical methods employed would allow this difference to be detected very easily; their sensibility would even enable us to discover a difference a hundred times smaller, corresponding to a velocity of translation ten times smaller than the velocity of the earth in its orbit.

67. Aberration of the Fixed Stars.

The negative result of the Michelson-Morley experiment was all the more unexpected, paradoxical even, seeing that, when applied to certain other optical phenomena, the elementary and apparently reasonable kinematic arguments which are used yield correct results. One of the best known of these phenomena is the *aberration of light*, or change produced in the apparent direction of the rays coming from the stars in consequence of the motion of the earth. The movement of rotation of the earth round its own axis, the velocity of which is much smaller [1] than the velocity of the centre of the earth in its rotation round the sun, may be regarded as quite insignificant in comparison with the latter. We shall, therefore, ignore the diurnal aberration, and consider the annual aberration only. If we take first of all a star situated at the pole of the ecliptic, it is clear from considerations of symmetry that it will describe a curve which is nearly a circle (since the earth itself describes round the sun an ellipse which differs but little from a circle). Indeed, the velocity of the earth being about 1/10,000 of the velocity of light, the angle between the apparent direction of the star and its real direction will be about 1/10,000 of a radian, which is equivalent to 20·62 seconds of arc (observation gives 20·47 seconds, which corresponds to a ratio of velocities slightly less than 1/10,000). Thus the apparent direction of the star will describe in one year a *cone*, half the vertical angle of which will be about 20 seconds; the star itself will describe on the celestial sphere a circle of radius equal to 20 seconds of arc. If the star is not at the pole of the ecliptic, we easily see that it will describe an ellipse (which reduces to a straight

[1] At the equator it is a little less than 500 metres per second, while the velocity of the earth in its orbit is about 30 kilometres per second.

line if the star is situated in the plane of the ecliptic), the semi-axis of the ellipse being always equal to the constant of aberration, that is 20·47 seconds. The phenomenon of aberration is sometimes compared to what happens when a person with an umbrella is walking in a shower of rain which is falling vertically. If he stands still, he has to hold his umbrella vertically overhead, whilst when he is walking he must incline the umbrella a little forwards, and the more so the faster he walks. We see that in this case the ordinary kinematic comparison leads to exact results. In fact, Bradley in 1727 was led by the theory of aberration to a value for the velocity of light which for a long time remained one of the best.

68. **The Doppler-Fizeau Effect.**

A more recondite phenomenon in which also the properties of light are in agreement with an elementary kinematic argument, is the phenomenon of Doppler-Fizeau, frequently called Doppler's principle. The nature of the phenomenon, which is very easy to observe in the case of sound, is this. If we are standing on the platform of a railway station when a fast train is passing with its locomotive whistling, we find that the sound suddenly becomes flatter when the train passes the station, that is to say when it is moving away from the observer instead of coming towards him. If we suppose the speed of the train to be 34 metres per second,[1] that is 1/10 of the velocity with which sound travels in air, it follows that if the train is 340 metres away, it will take 1 second for the sound it emits to reach us. The train passes us 10 seconds later, that is to say, 9 seconds after the sound it emitted 10 seconds ago has reached us. During these 9 seconds our ear has perceived all the vibrations emitted by the whistle during 10 seconds. If the whistle emits 450 vibrations per second, our ear perceives

[1] This corresponds approximately to a speed of 120 kilometres an hour, a speed actually reached by some fast trains.

500, and the pitch of the sound is thus raised in the proportion of 10/9, that is by a minor tone. When the train is moving away from us, the ear perceives during 11 seconds the vibrations emitted in the course of 10 seconds, and the pitch of the sound is thus lowered in the proportion of 10/11. Supposing the speed of the train to be 1/9 instead of 1/10 of the velocity of sound, the pitch of the sound perceived would be a major tone above the real sound, or a minor tone below it, according as the train is approaching or receding. When the train passes the observer, he will therefore perceive a sudden lowering of pitch amounting to two tones, that is to say from *me* to *doh* of the ordinary musical scale.

This phenomenon, so easily observed and explained in the case of sound, must take place in the case of light also, with this difference, that the *pitch* of sound is replaced by the *colour* of light, or, more precisely, by the corresponding spectral ray. The Doppler-Fizeau effect will thus show itself by a shift of the spectral lines. Our eye is more sensitive to luminous intervals than our ear is to musical intervals. The sensibility of the eye is increased in a considerable ratio by the use of the spectroscope and the method of interference, by aid of which certain wave-lengths can be estimated with a precision of one in a million. In the musical scale this would correspond to about one hundred-thousandth of a tone, or 1/10,000 of a *comma*.

On the other hand, the velocity of light being about one million times greater than the velocity of sound, it is much more difficult to devise an actual experiment in which the velocity of the source would amount to an appreciable fraction of the velocity of light. But although it may be difficult to realize such an experiment, the observation of certain double stars enables us to see the Doppler-Fizeau phenomenon. Think of a double star, that is to say, two brilliant stars rotating round their common centre of inertia. If the plane of the orbit passes through our solar system, it will happen at times that the velocity of one of the stars will be bringing it nearer us, whilst the velocity of the other will be taking it farther away. At the end of half a revolution (half a year of the system) the reverse will be the case, and the star which was approaching us will now be receding, and vice versa.

If one of the stars is considerably larger than the other, then only the motion of the smaller will be appreciable. In any case, at regular intervals, in consequence of the Doppler-Fizeau effect, we will notice modifications in the spectra enabling us to determine the period of revolution of the double star and its radial velocity. By the same method we can study the radial velocities of other stars. These measurements, however, are based upon the hypothesis that the periods of emission do not depend on the gravitational field (nor on any other physical circumstances), an hypothesis which is contradicted by the theory of general relativity. We can also try to verify the Doppler-Fizeau principle by observation of the rays from different parts of the sun's disc, some of which are moving towards us and others away from us, in consequence of the movement of rotation of the sun round its own axis.

To sum up: although some of the verifications require discussion, on account of the complexity of causes liable to produce the same effect, there is no doubt on the whole that the Doppler-Fizeau principle has been satisfactorily verified. In this matter, then, everything happens with light as with sound; the wave-length depends on the relative velocity of the source of the waves and the instrument which measures them (or the senses which perceive them).

69. Fizeau's Experiment on Running Liquid.

There are other optical phenomena the kinematic explanation of which is more simple. One of these phenomena, which has been known for many years, having been discovered by Fizeau, is the partial drag of light by running liquid. The velocity of light in water is less than its velocity in air; does it increase when the water is in motion instead of being still? Fizeau made an experiment which has become famous and has been frequently discussed from the theoretical point of view. According to the result of this experiment, the velocity in running water is greater than the velocity in water at rest, but the difference of these velocities, instead of being equal, as one might have expected, to the velocity of the water, is less than this velocity. The classical theories of the nineteenth century, by means of supplementary hypotheses, succeeded in giving a reasonable account of Fizeau's result. As we shall see in Note I (p. 199), the kinematics of relativity gives an immediate and natural explanation.

70. **Phenomena shown by Double Stars.**

Another phenomenon, in which the kinematical hypotheses deduced from ordinary mechanics lead to singular results, is supplied by those double stars which form a system analogous to the system which earth and sun would have formed, had the earth been larger than it is and, moreover, as bright as the sun. Let us try to form a clear idea of the appearances which such a system, observed from a distance, would present, if light actually behaved like a projectile, the velocity of which is the sum of the velocity given to it by the cannon and the velocity of the cannon itself. In other words, when the cannon itself is in motion in any way, the projectile has always the same velocity relatively to the cannon, for instance 1000 metres per second, at the moment of starting. Let us suppose, for the sake of simplicity, that the sun is motionless. In the month of January, the earth has a velocity of 30 kilometres per second towards a certain point in the sky. It is in this direction that we shall assume our observer to be stationed, at such a distance that it will take 2500 years for the light of the sun to reach him. The light given out by the sun on the 1st of January of the year 1 will thus reach this observer on the 1st of January of the year 2501, assuming him to have a chronology similar to our own. But by supposition, the light coming from the earth has a velocity equal to the velocity of light *plus* the velocity of the earth in its orbit; it is consequently greater by 1/10,000 of its value than the velocity of light coming from the sun. The time occupied on the journey is, therefore, less by 1/10,000 part, that is by three months for 2500 years, and it is on the 1st of October of the year 2500 that the earth will be perceived in the place which it occupies on the 1st of January of the year 1. Let us now pass to the 1st of July of that year. The earth has then a velocity of 30 kilometres per second, in the direction away from the observer; light from the earth will travel more slowly and will be perceived by the observer only on the 1st of October of the year 2501. The observer will further, *at this same time*, perceive the earth both in the position which it occupies on the 1st of January of the year 2, and on the 1st of October of the year 1, the date on which the velocity of the earth is perpendicular to the direction of the observer and consequently has no appreciable influence upon the velocity of light in this direction. We see how singular these appearances would be. One could get a clearer and more detailed idea by a discussion, which we shall omit, involving nothing but easy calculations in which only simple trigonometrical functions are employed.

We ought to state that the distance of the earth from the sun (which is traversed by light in about 8 minutes), as viewed from a distance of 2500 " light years ", corresponds to an angular distance the value of which in radians is the quotient of 8 minutes by 2500 years, that is to say extremely minute; it corresponds to about 1/1000 of a second of arc. It is easy to see, with the help of the laws of Kepler, that the value which this angular distance must have, if the phenomenon spoken of is to be possible, will stand (when the mass of the sun is given) in inverse proportion to the distance of the earth from the sun, that is to say, will diminish if we substitute for the earth another more distant planet. If the major axis is four times as great the velocity of the planet in its orbit will only be half, and the year on the planet will be eight times as long. In order to be able to observe the same phenomenon, we shall thus have to place ourselves at a distance sixteen times farther away, corresponding to double the number of years, these years being eight times longer. As the major axis has been quadrupled, it is seen at this distance, which is sixteen times longer, at an angle of one-quarter the old value. On the other hand, if the distance from the earth to the sun did not change, but the mass of the sun became four times greater, then the velocity of the earth in its orbit would be twice as great and the year consequently only half as long. The distance necessary to observe the same effect would then be divided by four, and the angular distance from the earth to the sun multiplied by four for an observer at this distance.

In reality, none of the double stars, including the spectroscopic double stars, show the singular phenomena of which we have given a survey. We must, therefore, conclude that the velocity of light does not depend on the velocity of the luminous source.

CHAPTER VI

The Special Theory of Relativity

71. What the Special or Restricted Theory of Relativity is.

The theory of relativity (which is now called the *special theory of relativity*, in contradistinction to the general theory of relativity with which we shall deal briefly in the next chapter) was produced for the purpose of giving a rational and coherent explanation of the apparently inexplicable experiment of Michelson and Morley. The negative result of this experiment is not the only confirmation of the theory. The new kinematics to which the theory leads and in which the velocity of the propagation of light plays the part of a limit-velocity is confirmed by other experiments also, particularly by the experiment of Fizeau on the partial drag of light by running liquid. The most complete proof, however, of the theory results from the study of the laws of electrodynamics as a whole, and from the fact that the system of equations governing these laws is invariant for the group of Lorentz transformations (see Note I). Our purpose here is not to give full proofs of the theory of relativity, nor even a complete explanation of the theory, such an exposition not being possible without the aid of a complicated mathematical apparatus. The theory of relativity is above all a mathematical theory. We shall try, however, to give a clear account of such results as can be expressed in ordinary language, and particularly to show that such results are not so paradoxical as some opponents of the theory are in the habit of proclaiming. It must, however, be admitted, in justification of the critics, that the champions of the theory of relativity too often delight to bring forward those results of the theory which appear to them to be specially fitted to shock the common sense of people who take statements too literally. The theory of relativity is not the only example of a physical theory which becomes absurd when its logical consequences are pushed to their very limit. Langevin's man who remains eternally young by keeping himself in unceasing motion at staggering speed, belongs to the same category as Jeans' pot which turns to ice instead of beginning to boil when placed upon the fire. There is no doubt that this phenomenon of Jeans' cannot be regarded as impossible; it is only highly improbable. Nevertheless, were a pupil of Jeans to come and tell his master that he had actually observed the phenomenon, he certainly would not be believed, the probability of hallucination or trickery, or some such cause of error, being *a priori* infinitely greater than the probability of such an extraordinary departure from the average laws of thermodynamics. It is therefore more reasonable to attribute the story of the pupil to one of the former causes rather than to the last. Nor would Langevin hesitate to intern a young man who came and told him that he was born under Louis XIV, but had remained young on account of his having travelled extensively.

72. Acoustic Signals and the Wind.

In order to familiarize ourselves with the arguments based on the transmission of optical signals, it will not be superfluous to study first the case of acoustic signals transmitted in air, a medium more familiar to us than the ether, that unknown and problematical support of optical phenomena.

Sound is transmitted in air with a velocity of about 340 metres per second (this speed depends on the temperature of the air, which we assume to be constant). If the air is still, and we give a sound signal such as the whistle of a captive airship, then, at the end of a second, this signal will be perceived at all points of a sphere having its centre at the point where the signal is given, and a radius of 340 metres. If, on the other hand, the air, instead of being still, is being carried along slowly and smoothly at uniform speed, then the sound waves will also be carried with it. If we give the signal in a moving airship which is itself carried along by the general motion of the air, then, at the end of a second, the signal will be perceived at all points of a sphere the centre of which is the point where the whistle will be after a second, and not the point where it was when the signal was given. If various observers were to be stationed in moving airships, all being equally carried along by the uniform motion of the air, the distance between the observers and the whistle would remain invariable in time. Supposing this distance to be 340 metres, then the observers will perceive the sound a second after it had been emitted, no matter whether there be no wind or a regular wind of any speed. The observers therefore, will not be able, by making experiments on the velocity of sound, to find out whether there is or is not any wind. In point of fact, our atmosphere is actually carried along in this way by the motion of the earth round its own axis. To an observer stationed on the moon and watching the clouds or gigantic moving airships they would appear to be in rapid motion, their speed exceeding 400 metres per second at the equator; but experiments on the velocity of sound would be useless for the purpose of demonstrating this motion.

Let us now consider observers stationed on the earth; if the air is calm, there is nothing to change in what we have just said with regard to observers stationed in airships. If we imagine that a large number of observers are scattered over a plain, and suppose that when one of them makes a sharp sound signal (blows a whistle, say) all the rest, the moment they perceive the signal, raise their arms for an instant, then the raised arms would at every moment mark out a circle having as its centre the point where the sound originates. An observer in an airship would see this circle being marked out and growing larger, and he would thus be able to tell at what point the sound signal was emitted, this being the centre of the circle.[1] If a wind was blowing, the observer stationed in a captive airship would see the centre of the circle formed by the raised hands displaced with the velocity of the wind.

73. The Timing of Clocks by Means of Acoustic Signals.

Let us now try to get an idea of how observers stationed at different points might time their clocks by making use of sound signals (we assume the observers to be separated by obstacles which prevent them from moving about among themselves with a portable chronometer). The means which seems to us to be the simplest, as it requires knowledge neither of the exact distance between the observers, nor of the exact velocity of sound, is the following: the observer A will give a signal and mark the time on his clock, say noon; when the observer B receives the signal, he, in his turn, will give another signal and mark the time on his clock; A will then mark the time when he observes B's signal, let us say, for instance, 4 seconds past noon. If we now admit that the time required for the transmission of the signal from A to B is the same as that required for the transmission of the signal from B to A, it is clear that B must have perceived A's signal and given his own [2] in turn at 2 seconds past noon. Such is the time which B's clock ought to show if it

[1] Of course we are supposing here that the propagation of light is instantaneous; the time taken by light to travel the short distances in question is imperceptible to our senses, and may be neglected.

[2] We neglect here B's *personal equation*, that is to say the time which elapses between the moment when A's signal is observed and B's signal is given; this time is of the order of a tenth of a second if B is a man who presses an electric button the moment he hears; it may be shorter if B is an instrument (interrupter).

is to synchronize with that of A. Should, however, B's clock show 3 seconds past noon, then we must conclude that his clock is in advance of A's by 1 second. This conclusion is, however, based upon the assumption of the equality of the times of transmission from A to B and from B to A. We know that this hypothesis is correct in the event of there being no wind, but ceases to be exact when there is a wind, and in the latter case the method would lead us to an inaccurate result. We need not take up time discussing how, assuming a regular and invariable wind, we might devise a coherent system of conventions with the object of defining, at least to a first approximation, coincidence of time at two different places, on the understanding that sound signals are the only means of communication. We should have occasion to consider the calculation of distances by means of the average velocity of sound, and the geometry of the triangle to which this method would lead. These considerations would not be without their practical utility, if we were studying the influence of the wind upon methods of surveying by means of sound.

We must also observe that for the hypothesis of a regular wind we could substitute the hypothesis of a parallel displacement in still air of all the points, the distances of which we are studying. These points, for instance, may be situated on the deck of a vast steamer moving at constant speed. This will help us to understand how, if our measurements of space and time were made by means of sound signals, the apparent effect of motion would be to modify the definition of time and to put space out of shape.

74. The Specification of Motion by Means of Acoustic Signals.

We prefer, however, to dwell upon a more interesting case, that namely of two systems, one of which is in motion relative to the other. For the sake of simplicity, we shall confine ourselves to geometry of one dimension, that is to say we shall consider exclusively points situated on a straight road (or straight railway line). On this road a car is moving; it is equipped for sending out and receiving signals by means of which it can communicate with an observer stationed at a point A on the road. For the sake of simplification, we shall assume that the speed of the car is exactly 1/10 of the velocity of sound. If we assume the observers to possess chronometers enabling them to measure time with great exactness,[1] it will be easy for them to determine the speed of the car in terms of the velocity of sound.

Let us assume, to begin with, that the fixed observer A sends out successive signals at intervals of 1 second, and that the car replies as soon as it observes the signals. If at the moment of the first signal the car is at A, the answer and its reception by A will coincide with this first signal. When A gives his second signal, the car will have advanced a distance of 34 metres, and whilst the sound is travelling over these 34 metres, the car traverses another distance of 3·4 metres; while the sound is travelling these 3·4 metres, the car traverses again a distance of 0·34 metres, and so on; the successive times will be 1 second, 0·1 second, 0·01 second, &c. The signal will thus be received at 0·111... second after it is emitted, and 1·111... second after the first signal. The time required for the sound to travel back being the same, A will observe the answer to the second signal 0·222... second after he has given it, or 1·222... second after the reply to his first signal. In the same way, he will receive the reply to his third signal 1·222... second later, that is 2·44... seconds or $2\frac{4}{9}$ seconds after the first signal; and so on. The exact measurement of these times will enable him, by means of an inverse arithmetical calculation, to find out the speed of the car, which in the above argument was supposed to be known (equal to 1/10 of the velocity of sound).

[1] Actually, for such experiments we would use, not chronometers in the strict sense of the word, but paper rolls unwound by clockwork, on which the phenomena to be observed would be recorded automatically.

The same experiments could be made by the travellers in the car. If they send out a signal at the end of 1 second, they will be at a distance of 34 metres from the point A; it will take 0·1 second for this signal to reach A, and 0·1 second to come back[1] from A to the point where the car was when the signal was given, or altogether 0·2 second. In the meantime the car has traversed twice the distance of 3·4 metres, and it takes the sound 0·02 second to traverse this distance; the calculation continues in the same manner, and we see that the answer is received 0·222 second (or 2/9 of a second) after the emission of the signal. The result is the same as in the case where A gave the signal, but we must point out that this result is not obtained in the same way as the other. When it is A who is giving the signal, while the car C is repeating it, the times required for the signal to travel forwards and backwards from A to C and from C to A are both equal to 0·11 second, the sum being 0·22... second. In the case where it is the car C that gives the signal and A that repeats it, the same sum is obtained, consisting of the time required by the signal to travel from C to A = 0·1 second, and from A to C = 0·1222... second. An important consequence results from this with regard to the mutual regulation of the chronometers of A and C by means of the sound signals. Such regulation requires in principle the determination of two distinct unknown quantities, namely the ratio of the time units (that is to say of the rates of the chronometers), and the difference in the time origins (epochs of reckoning), both operations being reduced to the reading of the two chronometers at epochs which are considered to be the same. Each comparative reading supplies an equation, and two equations will be sufficient to determine the two unknown quantities. From what has been said, A being at rest and C in motion, the time required for a sound signal to travel from A to C is

[1] We assume that A replies instantly to the signal received by sending out another signal.

equal to the time required for the returning signal to travel from C to A, since C gives the return signal immediately on receiving the signal from A. Consequently, the hour marked on C's chronometer at the moment when C receives the signal from A and gives the return signal must be compared with the arithmetic mean of the hour marked on A's chronometer when A makes his signal, and the hour when he receives the return signal from C. If there be agreement, then the chronometers are correctly adjusted. If the signal given by A is emitted, say, at noon, C receives it at 1 second past noon, and A receives the return signal at 2 seconds past noon.

Let us now suppose that, the chronometers being in agreement, we try to effect the adjustment of times by an inverse operation, that is to say, suppose that C gives the first signal and that A returns it immediately on receiving it. As we have seen, if C sends his signal at noon, and A receives it at 0·1 second past noon, C will receive the return signal not as 0·2 second past noon, but at 0·222 . . . second past. The arithmetic mean of the times marked by C will consequently be 0·111 . . . second past noon, whilst A marks 0·1 second past noon, and there is consequently a discrepancy to the extent of 0·011 . . . . Now as the signal was given when the distance between C and A was 34 metres, that is to say 1 second after the moment when A and C coincided, there is a difference of 1/90 of a second for 1·1 second, or 1/99 of a second for 1 second. It is easy to see that this difference will remain in the same proportion when C moves farther away, that is to say, it will be twice as great at the end of 2 seconds, and so on.

We have obtained this result by assuming the calculation to be made on the understanding that C is at rest and that A is moving away from him. Surely, however, such a blunder could only be made through carelessness. The travellers in the car know full well that it is they who are moving, whilst A is at rest. The error, however, will appear less improbable if we observe that what is really important is not the motion or rest relative to the ground, but the motion and rest relative to the air in which the sound is being propagated. If we assume that there is a strong wind blowing steadily at the rate of 34 metres per second in the same direction as that in which the car is moving, then all happens as if the car were at rest and A in motion. In this case, it is the adjustment of the chronometers by means of signals coming from C and returning to him which will give the correct results, whilst there would be a relative error of 1/99 were we to make use of signals coming from A and returning to him.

The conclusion which we have reached is thus as follows: In order to be able to regulate two chronometers at A and C by means of sound signals, it is not sufficient to know the relative motion of A and C; we must also know the intensity and the direction of the wind.

Even supposing, however, that we have no other physical method at our disposal than observations of sound, we can yet easily deduce from these observations the velocity and the direction of the wind, since we know that this velocity is either added to or deducted from the velocity of sound, according as it is travelling in the same direction or in the opposite. It will, therefore, be sufficient to place in A (or in C) two mechanical listeners 1 metre apart, and to measure with precision the time required for sound to travel to and fro between the two points, and the difficulty which had presented itself will be solved with ease.

75. Luminous Signals, and Intuitive Kinematics.

Let us now try to get an idea of the differences introduced in the preceding discussion when for sound we substitute light, or electrical oscillations (the Hertzian waves of wireless telegraphy), which are propagated just like light with a velocity almost a million times greater than the velocity of sound in air. The experiments will obviously be more difficulty on account of this great velocity. It has in fact been found impossible hitherto to carry out by direct experiment the measurement of the times required for light to traverse, for instance, two lengths of 1 metre oriented differently at a given point. All that we have hitherto been able to measure (experiments of Michelson) are the differences between the times required to traverse a certain distance forwards and backwards, AB + BA, and another distance forwards and backwards, AC + CA. It is easy to convince oneself that if the kinematics of light were the same as the kinematics of sound, then the experiment would enable us to determine the direction and intensity of the *ether-drift*, just as a similar experiment made on sound would enable us to measure the direction and the intensity of the atmospheric wind.

The result of Michelson's experiment being negative, we must conclude that the velocity of light is the same in all directions for an observer attached to the earth. As the velocity of the earth relative to the sun varies according to the season of the year, we must consider it as extremely improbable that the hypothetical ether participates exactly in these variations of velocity. That would be to return to the geocentric hypothesis of Ptolemy, given up since Copernicus and Galilei. It follows that the experiment would give the same result on a system having a different motion; that is to say, the velocity of light measured by specified observers does not depend on the velocity of these observers, but is invariable.

It might be objected that this phenomenon is conceivably due to the fact that the luminous source participates in the motion of the observers; this would imply that the velocity of light is influenced by the velocity of the source emitting it, contrary to what happens in the case of sound. This conclusion, however, would not only stand in contradiction to the Doppler-Fizeau effect, observed in the stars, but would also lead to consequences in disagreement with the observations on double stars which were discussed at the end of the preceding chapter.

There are, however, certain kinematical properties which we are compelled to admit for every agent which propagates itself in space, and is capable of leaving a trace of its passage at any point, these properties being purely geometrical and independent of the physical nature of the agent.

Let us imagine a number of men stationed along a straight line on a road, and raising their hands when they receive a signal. This of course is an extremely rough analogy, especially of what takes place with light signals; it means that we disregard the personal equation, that is to say the time which elapses between the moment when the signal was perceived and that when the reaction was completed. We might, however, substitute for the men chemical reagents (photographic plates, or gaseous mixtures detonating under the action of light), with which the personal equation, being infinitely smaller, must be much more constant. The roughness of the analogy, however, matters little; what interests us is the material representation of the passage of the ray or of the wave.

If successive signals be sent at regular intervals, then our men will also raise their hands at regular intervals, and we can note these intervals on our chronometer if we are standing near one of the men. But if we were travelling along the road in a car at great speed, in the opposite direction to the wave of light, it is clear that we should gain a little time in the interval between two signals, because at the moment of the second signal we would be beside a man who is nearer to the origin of the signals than we were at the moment of the first signal. This shows that the arguments with regard to sound are partly applicable to light, whilst the experiment of Michelson proves them to be partly inapplicable.

76. We must escape the Contradiction.

We thus arrive at a logical contradiction which it seems impossible to escape if we admit the absolute value of the definitions of space and time. Indeed, for the travellers in a car in motion on the road on which the men raising their hands are stationed, the velocity of light will be increased if they measure it indirectly by observation of the men, but must remain constant if they measure it directly by means of instruments which they are carrying with them. As velocity is the quotient of a distance by a time, we can escape the contradiction only by modifying the value of the distance or the value of the time, or the value of both simultaneously, in such a manner that the values are not the same in the two contradictory experiments. Fitzgerald and Lorentz have proposed to explain the result of Michelson's experiment by postulating that the measurement of space is not the same for observers in motion as for observers at rest. This is the hypothesis of the Lorentz-Fitzgerald contraction, according to which lengths are shortened by motion. This hypothesis led to a system of formulæ bearing the name of the Lorentz transformation, and this system, in its turn, suggested to Einstein an explanation which not only explains Michelson's experiment, but also squares admirably with several other physical phenomena hitherto unexplained or only badly explained.

77. The Independence of Space and Time.

The essence of Einstein's hypothesis, which is known as the special theory of relativity, consists in the postulate or axiom that there is no reason whatever for assuming *a priori* that space and time can be defined independently of one another, that is to say independently of motion. In other words, given two physical phenomena, such as two gaseous explosions, one on the sun and the other upon earth, there is no sense *a priori* in saying that these phenomena are simultaneous, since simultaneity cannot be defined unless some physical means be given by which it can be tested experimentally. If we admit these premises, we find first of all that, in the actual state of our knowledge, the only means accessible to us by which we can try to define simultaneity at two distant points is the employment of light signals. We then find that the use of these leads to different definitions according as the experiment is made by one set of observers or by another set who are in motion relative to the first. The definition of time, or rather of simultaneity at the various points of space, is thus relative, not being the same for different groups of observers. The relativity of time involves the relativity of the measurement of space. Let us consider two points of a ruler which we have beside us; to anyone who believes in absolute space, the distance between these points means the distance between two points of absolute space with which these two points of the ruler coincide at one and the same instant; for the ruler is carried on by the motion of the earth, and each of its points, if absolute space be defined by axes connected with the sun, is displaced with a velocity of 30 kilometres per second. If, therefore, for an observer in motion relative to us, the definition of simultaneity for the two points of the ruler is not the same as our definition of simultaneity, then this observer will not measure the distance between the same points as we have chosen, and the length of the ruler will be different for him.

78. The Special Theory a Logical Consequence of the above Premises.

The reader who thoroughly understands the foregoing remarks may be said to have taken the most difficult step on the road leading to the special theory of relativity, what is left being nothing but a purely mathematical question offering no particular difficulty. Those who are not mathematicians may accept the results of these deductions and calculations with the confidence which they usually repose in mathematicians. The premises once admitted, no new postulate is introduced, and no contradiction need consequently be feared, since no error in either argument or calculation is made.

If the consequences sometimes appear paradoxical, this is due to the fact of our forgetting the premises; we forget, that is to say, that we should never speak of the date of an event without specifying the group of observers relative to whom time is defined.

Besides, the special theory of relativity is strictly applicable only to groups of observers whose relative motion is rectilinear and uniform. But every relative motion of two sufficiently small portions of space may be considered as being rectilinear and uniform, if it be continuous and contemplated during a sufficiently short space of time, just as every arc of a curve over a small length can be identified with its tangent, or every surface with its tangent plane.

79. Examination of an Objection.

One of the most interesting consequences of the restricted theory of relativity is the quite special part which it assigns to the velocity of light, making it appear as a velocity which cannot be exceeded. To people accustomed to handle the formulæ of mechanics and of classical kinematics and to reflect upon the meaning of these formulæ, this is the consequence which will, perhaps, appear as the strangest and the most difficult to admit. The classical law of the composition of velocities appears indeed to be so intuitively evident for the small velocities which we can observe that there seems to be no way of escaping it. If a moving footway has a speed of 6 kilometres per hour, and I walk upon this trackway at 6 kilometres per hour, I shall make 12 kilometres per hour when walking in the direction of motion of the footway, but shall remain at one spot if I walk in the opposite direction. That is quite evident; but if I apply the same argument to a ray of light, the theory of relativity disappears.

To this objection, which, so to speak, always turns up again no matter how conclusively we imagine we have disposed of it, it is easy to reply that the theory of relativity had its origin precisely in the fact that experiment had flatly contradicted this apparently irrefutable argument. Apart from the Michelson experiment, of which we have already spoken, it may be worth while also to refer to the famous experiment of Fizeau on the velocity of light in running liquid (see Note I).

80. The Possibility of Continual Increase of a Velocity does not Involve the Conclusion that the Velocity may Increase Indefinitely.

It appears, however, that we might raise another objection to this notion of a maximum velocity that cannot be exceeded, an objection of a dynamical and not of a kinematical character. The action of a force, or of a blow in a proper direction upon a body in motion, will result in the speed being increased; if then we start with any velocity, this velocity can always be increased, so that there can be no velocity too high to be reached. The objection, presented in this way, contains in the first place a flaw which will not escape the notice of the reader who has even a slight acquaintance with the use, if not with the theory, of numerical series. Such a series may quite easily consist of an infinity of positive terms and yet be convergent, that is to say have a *finite* sum. One of the simplest instances is the series in which every term is one-half of the preceding term:

$$1 + 1/2 + 1/4 + 1/8 + 1/16 \ldots .$$

This series, the sum of which is 2, was known to the Greeks and utilized in the famous sophism of Zeno of Elea. It can thus be easily imagined that the speed of a body may constantly increase so as to approximate indefinitely closely to a certain limit, viz. the velocity c of light, without ever attaining this limit; this, however, involves the consequence that the increase of the velocity necessarily becomes very small, however strong the acting force may be, when the velocity of a body comes very near the velocity of light. We can translate this fact into the language of ordinary mechanics by saying that the mass of a body increases with its speed and tends towards infinity as this speed tends towards c; the resistance offered by the body to an increase of velocity, that is its inertia, depending in fact on this mass. The velocity being relative, since distance is relative, the mass also is relative to the system of comparison chosen. All this is perfectly coherent, and the formulæ which translate into mathematical language all that we have just expressed in ordinary language cannot lead to any contradiction (see Note I).

81. Instantaneous Propagation has as Little Plausibility as a Velocity that cannot be Exceeded.

It would be useless to attempt to disguise the physical hypothesis which lies at the root of the theory of relativity. According to this hypothesis, not only *do we not know* of any velocity exceeding that of light, but *there can never be* any such velocity. To certain minds this hypothesis appears not only improbable but even absurd, and one must admit that it is at least incompatible with our traditional ideas about space and time. In fact, as soon as we admit the absolute conceptions of time and space, the theorem of kinematics relative to the composition of velocities follows logically. If a train is moving relative to us with a speed v, and if a traveller is walking in the train with a speed v' relative to the train, then the speed of this traveller relative to us is $v + v'$. It is therefore possible to realize, or bring into concrete existence, the velocity $v + v'$, since we know how to realize, on the one hand, the velocity v, and, on the other hand, the velocity v'. If in particular we take v' to be equal to v, then, from the possibility of realizing the velocity v, we can infer the possibility of realizing the velocity $2v$; we can then pass to the velocity $4v$, then to $8v$, $16v$, $32v$, $64v$, &c. There is nothing to stop the mathematician in these successive duplications, since the premises from which we start are not disputed.

Setting aside for a moment the experimental reasons for replacing the law of classical mechanics on the composition of velocities by another law, in which the velocity of light definitely intervenes as a limiting velocity, let us consider the matter from the standpoint of the *a priori* arguments, often based on sentiment rather than on logic, which critics of the theory sometimes adopt. We find ourselves reduced to a choice of two alternatives: either we must accept, or we must deny, the absolute validity of the argument that, whenever an assigned velocity is admitted to be possible, then double that velocity must be admitted to be equally possible. If we accept this argument as valid, we see that there is no limit to possible velocities, and conclude that there must be actions whose velocity of transmission may exceed any number given beforehand, that is to say, be infinite in the exact sense attributed to this term by mathematicians. We know in fact that mathematicians do not deal with the actually infinite (except in certain portions of the theory of aggregates), but give the name infinite to a variable number, the value of which may become and remain superior to any number fixed beforehand. Now the instantaneous transmission of any action to a distance is a conception to which scientific and philosophical minds have always been averse; when this conception was accepted as a device of calculation (as in the case of gravitation in celestial mechanics) astronomers were under no illusion with regard to the real basis of the hypothesis.[1] Physical theories

[1] I may be allowed to recall that in the preface to my book *Le Hasard*, written in 1914, I do not hesitate to characterize as *absurd* the classical statement of Newton's Law, according to which gravitation is trans-

have always had a tendency to define a medium by which actions are transmitted to a distance; in other words, a tendency to substitute for action at a distance the mutual reaction of elements immediately adjacent to each other, this reaction being propagated with a certain velocity, necessarily finite. The argument of common sense, according to which every velocity may be doubled, thus clashes with the other argument of common sense which says that there must be a limit to the rapidity of propagation if we refuse to admit the possibility of instantaneous propagation.[1] A little reflection will thus suffice to discount considerably the force of this so-called argument of common sense, and to commend the conception of an upper limit to velocities which are possible as a perfectly plausible *a priori* conception. But the argument then takes a new turn, leading us to deny completely the validity of traditional kinematics and to assert the necessity of substituting for it another kinematics which will respect our natural inclination to deny the possibility of instantaneous action at a distance, that is to say, which will assign a superior limit to possible velocities. Now the kinematics of restricted relativity is evidently, if not the only type, at least the simplest type of kinematics which we can devise *a priori* so as to satisfy these conditions.

In the formulæ of this kinematics there figures a certain velocity c which cannot be exceeded; experiment confirms these formulæ, on the understanding that it is just this velocity c which is to be identified with the velocity of light.

The physical hypothesis to which we have been led is

mitted instantaneously. It is sometimes said that this instantaneous transmission follows from the work of Laplace. The matter is not so simple, however; Volume IV of Tisserand's *Traité de Mécanique Céleste* may be consulted on the subject.

[1] When we admit that energy is transmitted with finite velocity, we ought not to shrink from the deduction that this energy is for part of the time localized in the transmitting medium. But we should be drawn too far if we began to discuss the question of what becomes of the conservation of energy when we accept the relativity of time.

thus quite as plausible as many other hypotheses which have gained a firm footing in science. We cannot maintain, of course, that the hypothesis will never be modified by any future development of science, but, for the moment, it is among the hypotheses which must serve as the starting-points for any such development.

82. Spatial Measurement of Time: Einstein's Interval.

Let us now try and sum up in ordinary language the conception of space and time to which the special theory of relativity leads us.[1] The essential point is the absolute impossibility of separating the measurement of space from the measurement of time. For the purposes, however, of science and practice, we must come to definitions which are as precise as possible and can be expressed in numbers, and we shall therefore adopt the definition of time which is accepted by the astronomers. But we must not forget that this definition has no absolute value, being relative notably to the fact that it has been established by observers attached to the terrestrial globe (or, more generally, to the solar system). In a similar way, we habitually make use of the vertical as a privileged direction for the purpose of specifying position on the earth's surface, although we have known for a long time that the earth is in motion in space and that the vertical is of no more importance in geometry than any other direction.

This applies also to space. All that we have said with regard to the methods of successive approximation by means of which we have succeeded in making a survey of the terrestrial globe, and even of the stars, retains its value. By the classical methods we obtain a first approximation which is not only sufficient in almost every case, but which

[1] For the mathematical presentation of this theory reference may be made to my *Introduction mathématique à quelques théories physiques* (Paris, Gauthier-Villars, 1914).

remains the necessary starting-point for subsequent approximations.

It is only when we have these new approximations in view, with reference to certain phenomena of which we wish to make a more minute analysis, that it will be necessary to take into account the relative nature of the measurements of space and time. We shall have to combine these measurements with each other, without, however, forgetting the elementary arithmetical rule according to which none but quantities of the same nature should be added together. In order to reduce time and space to a common measure, we shall make use of the velocity of light, the privileged rôle of which we have recognized. For a certain observer, a given spatial distance is traversed by light in a definite interval of time, and we shall agree to the convention that for this observer the spatial distance and the time required to traverse it are equivalent. If the time is 1 second, the corresponding distance in space will thus be 300,000 kilometres. It is in this sense that we may say that the theory of relativity leads us to consider time as a fourth dimension of space. This simply means that, thanks to the intervention of the velocity of light, every time-interval may be measured in length units, the spatial value of a time-interval being equal to the distance traversed by light during this time-interval.

The advantage of this spatial measurement of time would be small, but for the fact that it enables us to dispense with the consideration of a particular observer. We have said that the definitions of length and time are not the same for groups of different observers in motion relative to each other. Let us suppose that several groups of observers observe the same two events, for example the collision of two bodies, and the explosion of a mass of gas. For the diverse groups of observers, the distances of these events in space will be different, and the distances in time equally different. If, however, we estimate the distance in time in terms of space units, and if we call the *interval* of two events the length, the square of which is equal to the difference between the square of their distance in time and the square of their distance in space, then the value of this interval will be the same for every group of observers. Here we have a physical fact which in the present state of science it seems impossible to dispute. It follows that there is a very great advantage in making use exclusively of formulæ containing *intervals* and not separate distances in time and space. This advantage is quite of the same order as the simplification introduced into mechanics by the use of Galilean axes. It is, therefore, *convenient* to use intervals, just as it is *convenient* to assume that the earth is rotating and the sun standing still (to a first approximation). Moreover, we must not forget that for Poincaré this *convenience* is identical with scientific truth.

83. The Principle of Causality is not at Stake.

The distances in time of two events are generally greater than their distances in space (let us not forget that 1 second of time equals 300,000 kilometres); the square of the interval we have defined will then be a positive number, the calculation of this interval offering no difficulty whatever. Such is not the case when the distance in time is less than the distance in space. In this case, the square of the interval of two events being a negative number, the rules of algebraical calculations lead us to consider this interval as a purely imaginary quantity. There is nothing in this formal result to shock or even surprise a mathematician, who is accustomed to see imaginary quantities applied with advantage to the study of perfectly real phenomena. It will not be superfluous, however, to point out that no mystical metaphysical significance need be attached to the fact that the interval presents itself here under the special algebraical form of a quantity whose square is negative.

This particular case is nevertheless very interesting, for we can easily show that when two events are, for one group of observers, thus nearer to each other in time than they are in space, then not only is this the case for every group of observers (this being a simple consequence of the invariability of the interval, the square of which must remain negative), but, moreover, we can find groups of observers for whom the distance in time of events considered in a given order will be, at our choice, either positive, nil, or negative. The event A, according as we identify ourselves with this or that group of observers, may thus be considered as being either anterior or subsequent to event B, or, as a special case, as being simultaneous with B. This inversion of the order of time has been held to involve an infringement of the logical laws of causality. If A be at once anterior and subsequent to B, may it not also be considered as being both the cause and the effect of B, which would imply a contradiction? In reality, if we admit the physical postulate of the impossibility of a velocity being greater than the velocity of light, an event A cannot be the cause of an event B, unless it precedes it by an interval of time sufficient for some action to be transmitted from A to B. In the case where the distance in space exceeds the distance in time, this transmission is impossible, the distance being too great or the time too short for A to influence B, or for B to influence A, so that neither of them can be the cause of the other. For observers whose motion is such that the distance of the events in space is the least possible, the events appear to be simultaneous.

If, on the contrary, the distance in space is less than the distance in time, then we shall reduce this distance in time to a minimum by choosing a group of observers for whom the events will appear to coincide in space. There is nothing to astonish us in the fact that two events A and B (subject only to the restriction that the square of their interval is positive), can be considered as coinciding in space, it being sufficient if the observers have time to pass from A to B between the moment when A takes place and the moment when B takes place. If these observers were to imagine themselves to be at rest, and saw A and B occurring beside them, they would conclude A and B to be occurring at the same point in space. It is thus that we habitually judge of two terrestrial events. Two events occurring in the room where we are, at an interval of 1 second, are considered by us to be occurring at the same place. For an observer stationed on the sun, the distance in space of these two events would be 30 kilometres, and for an observer stationed on the moon it would be several hundred metres (about 500 metres, if our room were situated at the terrestrial equator).

84. Restricted Relativity concerns only Translations.

In conclusion, let us repeat that the special theory of relativity is applied only to observers whose relative motion is one of uniform translation; we must also assume the directions of their axes to be fixed with respect to the stars. But just as every sufficiently small arc of any continuous curve may be considered to be rectilinear, so every continuous motion, contemplated during a sufficiently short space of time and in a sufficiently restricted portion of space, may be considered as a uniform motion of translation. It is owing to this fact that we have been able to speak of the aspect of phenomena occurring at a second's interval for observers stationed either on the earth, the moon, or the sun. The case, of course, is different when we contemplate a motion for a sufficiently long time, or when we simultaneously consider the motion of different points of such a mass as that of the terrestrial globe, even for an infinitely short period. Thus the special or restricted theory of relativity by no means permits us to say that it is indifferent whether we consider the earth to be at rest or to be turning round its own axis and round the sun; it does not at all modify the conclusions of Copernicus and Galilei.[1]

[1] To the works already cited we must add a substantial opuscule by M. Émile Picard, *La Théorie de la relativité et ses applications à l'astronomie* (Gauthier-Villars). We may also point out here that we may try to escape the contradiction referred to on p. 154 by assuming that we can define in space two different velocities of light, the velocity of energy and the velocity of vibratory movement, and that, according to the experimental method adopted, we measure the one or the other of these two velocities. It appears that the objections raised by M. Sagnac to Einstein's theory are related to this point of view, but that a sound experimental and theoretical basis for them is still lacking.

CHAPTER VII

The General Theory of Relativity

85. The General Theory of Relativity is above all a Mathematical Theory.

The general theory of relativity is beyond the scope of the present work, but I can scarcely conclude without at least a few remarks on the subject. I shall be obliged, however, to employ mathematical formulæ, a means of discussion which I have succeeded in avoiding up to this point, and which will no doubt repel the majority of my readers. On the other hand, readers to whom these formulæ are no obstacle will find them decidedly incomplete; it is to more technical works that those readers will turn. In short, the following pages are meant for those who will only half understand them, in the hope that they may stimulate curiosity and induce some to acquire the indispensable mathematical knowledge without which the general theory of relativity cannot be approached to any good purpose.

86. Euclidean Geometry and Curvilinear Co-ordinates on Surfaces.

Euclidean geometry is dominated by the theorem of Pythagoras, which is expressed in the formula

$$a^2 = b^2 + c^2, \dots\dots\dots\dots\dots (1)$$

a being the hypothenuse of a right-angled triangle, and b and c the lengths of the two sides containing the right angle. If, using the notation of the differential calculus, we denote by dx and dy two infinitely small lengths at right angles to each other, and by ds the hypothenuse of the right-angled triangle which they form, then this formula will become:

$$ds^2 = dx^2 + dy^2. \qquad (2)$$

Under this form it represents the element of the arc, ds, in rectangular co-ordinates, x, y; the length of any arc of a curve is expressed by the integral:

$$\int ds = \int \sqrt{dx^2 + dy^2}. \qquad (3)$$

Let us now place ourselves, not upon a plane but upon a curved surface, of any form whatever, but continuous. In the neighbourhood of a point O, the surface will sensibly coincide with its tangent plane, and if we denote by Ox, Oy two directions at right angles in the tangent plane, the formula (2) will hold good. However, when the point O changes its position on the surface, it will not in general be possible to define two continuous variables x and y for the whole surface, such that the formula (2) is always true, so that we can infer the formula (3); this will only be possible when the surface is developable, or, in other words, applicable upon a plane. On surfaces which are not of this type, we can define curvilinear co-ordinates u, v, that is to say, a network of curves analogous to the network formed in the plane by parallels to the axes Ox, Oy; the formula (2) is then replaced by the following:

$$ds^2 = \mathrm{E}du^2 + 2\mathrm{F}dudv + \mathrm{G}dv^2, \qquad (4)$$

in which E, F, G are *a priori* any functions whatever [1] of u and v. At any given point, these functions take numerical values E_0, F_0, G_0 and the formula (4) becomes

[1] It is always permissible to suppose the co-ordinates rectangular, that is to say, to take F = 0; but this special hypothesis leads to no real simplification in our exposition, and is rather a hindrance to generalization.

$$ds^2 = E_0du^2 + 2F_0dudv + G_0dv^2$$
$$= \left(\sqrt{E_0}du + \frac{F_0}{\sqrt{E_0}}dv\right)^2 + \left(\frac{\sqrt{E_0G_0 - F_0^2}}{\sqrt{E_0}}dv\right)^2, \quad (5)$$

and takes the form (2) if we put

$$x = \sqrt{E_0}u + \frac{F_0}{\sqrt{E_0}}v,$$
$$y = \frac{\sqrt{E_0G_0 - F_0^2}}{\sqrt{E_0}}v. \quad \ldots\ldots\ldots\ldots(6)$$

Thus we see that the general linear element (4) summarizes and integrates, as it were, the properties of an infinity of elementary plane elements of the form (2), the variables x, y having different meanings for each of them, meanings defined by formulæ (6).

The various surfaces which have their linear element defined by the formula (4) have, moreover, on the one hand different properties, and on the other hand common properties. These common properties are called intrinsic properties of the linear element (4); they can be defined by the invariants of this linear element, invariants of which the most important is the total curvature (we may mention here also the geodesic curvature).[1]

87. The Interval generalized by Means of the Quadratic Form in Four Variables.

Just as Euclidean geometry is dominated by Pythagoras' Theorem, so the special theory of relativity is dominated by the formula for the interval between two events, which may be written.

$$L^2 = c^2T^2 - D^2, \quad \ldots\ldots\ldots\ldots(7)$$

L being the interval, cT the distance in time, D the distance in space; T is the measure of the time between the events,

[1] See Darboux, *Leçons sur la théorie générale des surfaces.*

estimated in time units; cT is the same measure estimated in length units, c being the velocity of light. If we denote by ds the *elementary interval* between two events infinitely near each other both in space and in time, by dt the elementary temporal distance in time units, and by dx, dy, dz the rectangular components of the infinitely small spatial distance, the formula (7) becomes

$$ds^2 = c^2dt^2 - dx^2 - dy^2 - dz^2. \quad \ldots\ldots(8)$$

This formula holds at every point of the universe, that is to say, at every point of space at any moment of time; but, if we change our position in space and in time, we cannot define co-ordinates x, y, z, t, the same throughout, such that this formula always holds. Just as any curved surface possesses at every point a tangent plane, so the real universe possesses at every point a tangent universe for which formula (8) holds good, and which we may call the tangent Lorentz-Minkowski universe, just as the tangent plane to a curved surface is a Euclidean plane.

For formula (8) we are thus led to substitute a formula the second member of which is a general quadratic form in four variables:

$$\left.\begin{aligned} ds^2 = g_{11}du_1{}^2 + g_{22}du_2{}^2 + g_{33}du_3{}^2 + g_{44}du_4{}^2 \\ + 2g_{12}du_1du_2 + 2g_{13}du_1du_3 + 2g_{14}du_1du_4 \\ + 2g_{34}du_3du_4 + 2g_{42}du_4du_2 + 2g_{23}du_2du_3 \end{aligned}\right\}, \quad (9)$$

where the g's are functions, *a priori* unrestricted, of the four variables u_1, u_2, u_3, u_4.

If we assign to these variables definite values, the quadratic form (9) will have its coefficients constant; it can therefore be reduced to the form (8) by a suitable change of variables; we shall assume that the values of the g's do actually allow this reduction to be made by a real change of variables, that is to say that we can find a positive square c^2dt^2 and three negative squares $-dx^2$, $-dy^2$, $-dz^2$. The separation of space from time is thus effected.

We shall assume that the physical properties of the universe are exclusively intrinsic properties of the linear element (9), that is to say, that they can be expressed by means of invariants of this element, and so in a form which remains invariant for every change of variables effected on u_1, u_2, u_3, u_4. The assumption thus stated constitutes the principle of general relativity.

H. Weyl has recognized the necessity of adjoining to the quadratic form (9) a linear differential form,

$$dl = - l(\phi_1 du_1 + \phi_2 du_2 + \phi_3 du_3 + \phi_4 du_4), \quad (10)$$

intended to define at each point the units of measurement or, if we prefer to say so, the metric of the universe. This theoretical conception appears to me to involve some debatable points (see Note II); what is important about it is the precise physical interpretation we are led to give to the system of invariants of the forms (9) and (10); the laws of physics, electromagnetic phenomena as well as gravitation, are contained in these invariants.

A few words only with regard to gravitation. We have known for a long time how to put the general equations of mechanics in such a form that they express a certain minimum property relative to a quadratic form. In other words, the general problem of mechanics is tantamount to the finding of the geodesics of a certain quadratic form. A change of variables referring to space and time transforms a problem of mechanics into a problem of relative motion, to which the same principles are applicable. The essential point in Einstein's theory of gravitation is the idea that a field of gravitation does not differ from a field of forces of inertia. When we are falling freely, that is to say when we are yielding to forces of gravitation (the bullet of Jules Verne), everything takes place as if these forces did not exist. Thus had the earth's crust been sufficiently elastic to yield completely to the attraction of the sun and the moon (tides), then the diurnal variations of this attraction could not have been disclosed by the pendulum, any more than the total attraction of the sun, in respect to which the earth is falling freely, since it is moving freely in space. If, conversely, we were to communicate to a body an acceleration which is the reverse of that which a field of gravitation would produce, everything happens as if the body were submitted to this field. Our apparent weight increases in a lift which is beginning to rise rapidly, and diminishes when the lift comes down with a speed which is rapidly accelerated. All this is a consequence of the well-known fact of the rigorous identity of the Newtonian mass (the gravitational, or attracting and attracted mass) and the Galilean mass (or inertial mass). It is thus possible in the equations to replace the presence and action of matter by modifications equivalent to the displacements produced by forces of inertia. In mathematical language this amounts to saying that matter manifests itself solely by deformations of the four-dimensional space-time given by formula (9).

88. Change of Variables in Mathematical Theories.

I must confine myself to these brief indications, which could not be developed further without the help of a considerable mathematical apparatus. I should, however, like to say a word on one point which does not seem to me to have been sufficiently elucidated in the expositions of the general theory of relativity and in the discussions to which they have given rise. It is a well-known fact, brought out clearly by Poincaré, that any physical theory admits of an infinity of mathematical translations, all equivalent, and deducible from each other by a change of variables, or, as we may say, of notation. These mathematical theories, however, although equivalent, are far from being all equally simple, and there is an advantage in choosing the simplest of them and, at the same time, if possible, those in which the transition between the formulæ and the physical observations is the easiest. If, however, we put aside this criterion of simplicity, we can say that among all the mathematical theories, deducible from each other by means of a change of variables, none is worth either more or less than the rest. It is therefore natural to ask whether we are really obliged to make a choice among these equivalent theories and whether we might not retain them all, or, rather, retain that which they have all in common, namely, their invariants; it should then be possible to express all natural laws in terms of these invariants. If we were to stop here, we should only have stated a mathematical truth, indisputable indeed, but for that very reason sterile, a truth which by itself could not help us to make any progress in the knowledge of the laws of nature. In fact, we ignore everything, both as to the number of the equations and the number of the variables figuring in them, and it is thus impossible even to approach the question of the form of these equations. The general theory of relativity consists in the affirmation that the equations (9) and (10) include precisely the number of variables and of functions necessary and sufficient for the definition of the totality of the phenomena and their laws. These equations are transformed into equations of the same form by a general point transformation in four variables, and it is the invariants of this system of equations which enable us to express the laws of the universe.

89. Can a Few Equations contain the Geometrical Universe?

The claim that we can include in a few coefficients the infinite variety of the perceptible world with its boundless diversity of qualities may appear extraordinary. But it will scarcely surprise the mathematician, who knows all that the apparently simple notion of an arbitrary function contains; it may not be superfluous to dwell a little on this point. We know that in order to define a surface in

Cartesian co-ordinates, it is sufficient to give the equation

$$f(x, y, z) = 0 \ldots\ldots\ldots\ldots\ldots(11)$$

between the three rectangular co-ordinates x, y, z; such an equation being able, with a suitable choice of the function $f(x, y, z)$, to represent any surface whatever, of a form as strange and complicated as we may wish. It may, therefore, if we take the point of view of *form* alone, represent the entire universe, that is to say all the surface boundaries separating any two different substances (here for a moment we are assuming the hypothesis that all motions could suddenly be stopped and the world remain absolutely motionless; in other words, we are considering what would have been called in the nineteenth century the universe at a determinate moment $t = t_0$). No mathematician has, of course, ever thought of actually calculating, we do not say the function $f(x, y, z)$ which would represent the entire universe, but even simply the functions f which would exactly define the Venus of Milo or the surface of a stormy sea. There is no doubt, however, that if we assume the Venus of Milo or the sea to possess, at a given moment, a definite geometrical form, then it would be possible to represent these forms by the methods of analytical geometry, and they would consequently lead to such equations as (11). We do not imply that the mathematician capable of studying certain general properties of equations like that of (11), would have been also capable of imagining and producing the Venus of Milo: we shall not even suggest that algebraical equations can be of any use in the study of antique sculpture, but we nevertheless can state positively that all forms, either realized or realizable by artistic genius, simply on account of the fact that they are constructed in three-dimensional space, can be expressed in equations, such as (11). It would be child's play, moreover, for the analyst to combine into a single equation all the equations corresponding to the different surfaces by means of which a geometrical description of the world could actually be given; but of course such a unique equation, although easy to write in the form of (11), would be absolutely unmanageable and useless if our object was to bring out the details. Must we, therefore, conclude that there is no interest, perhaps even no sense in saying that the equation (11) could at least be imagined, if not written out explicitly? I do not think so; for one thing we thus express a fundamental fact, namely, that we consider space to be a continuum of three dimensions, so that we can connect the properties of any particular figure extended in space with the properties of (continuous) functions of three real variables.

90. Is the World Simple?

We can see now, I think, the justice of the remark made in the concluding section of our Introduction. Thanks to the theory of Einstein, the famous law of Newton, instead of remaining isolated in science, takes its place within the framework of a more general theory, which embraces, in particular and notably, optical and electromagnetic phenomena. No scientist will deny that such a reduction of widely differing phenomena to common principles, when found possible, constitutes an advance. Science has always endeavoured to give an explanation, or rather a description, of the world by means of the smallest possible number of simple elements. The nineteenth century was the century of mechanical explanations; the twentieth century will, perhaps, be the century of geometrical explanations. Is this tendency of science legitimate? Is it not based upon a postulate that cannot be proved, namely, that of the simplicity of the world? And if this postulate is false, are we not following a wrong track and getting farther away from the truth instead of coming nearer to it when we try to bring into mutual dependence phenomena which are essentially and inevitably distinct? Such are the objections which naturally present themselves. I am not pretending to settle in a few lines the philosophical question which they raise; I only wish to try to define the terms used, defining as exactly as possible what a simple scientific explanation of the world ought to be, and considering in what sense the theories of Einstein may claim to supply us with such an explanation.

There is no lack of arguments against the simplicity of the world; they have been set forth in a particularly striking manner in a fine work by M. J. H. Rosny aîné.[1] I do not believe that it would be possible to enter an abler or better attested plea against the idea that the world can be reduced to one or two simple elements. Leaving to others the task of discussing the philosophical theses of M. J. H. Rosny aîné, and taking up myself the scientific position exclusively, I shall try to show in what sense it is legitimate and even necessary to search for unity in the scientific explanation of the world.

91. The Virtuoso and the Phonograph.

There is no need to remind the reader of the fact that the aim of science is to understand and to predict phenomena, and that this aim can only be attained by an exact numerical description of these phenomena. For the scientist, an explanation of the world can only mean a numerically exact description of the world. If this numerical description could embrace the future as well as the past, the most exacting would declare themselves satisfied. Meteorology, for instance, would be sufficiently perfect if it allowed us to foretell with certainty, for any day of the coming year, the temperature, the pressure and hygrometrical state of the atmosphere, or the number of inches of rainfall and the direction and velocity of the wind at an assigned point on the earth's surface.

If we accept this point of view, we may be tempted to conclude immediately that number is the sole element

[1] J. H. Rosny aîné, *Les sciences et le pluralisme* (F. Alcan).

with the help of which the world can be completely described; this is the old Pythagorean thesis, which has never ceased to find followers. If, however, we look a little more closely, we shall find that number alone is not sufficient. Confining ourselves to the example of meteorology, we are able, it is true, by means of four numbers, to define temperature, pressure, and the direction and velocity of the wind. We may say, for example, that the temperature is 15° Centigrade, the pressure 762 millimetres of mercury, that the direction of the wind makes an angle of 65° with the north (towards the west), whilst the velocity of the wind is 10 metres per second. These four numbers, however, namely 15, 762, 65, and 10, have different significations, corresponding to distinct qualities and, at first sight, are entirely irreducible to each other. If the number of measurable quantities is unlimited, the apparent simplicity of number conceals a complexity to which we must resign ourselves unless we can reduce it; and in fact the reduction of apparently innumerable qualities to as small a number as possible of *essential* qualities is, perhaps, the most important task of science. But it must be understood that in this reduction the scientist takes into account only measurable properties and not those of a psychological character. Take the case of a violinist whose playing is moving his audience profoundly. The scientist does not analyse this emotion, or try to find out the part played by the genius of the composer or the talent of the player; but if he succeeds in producing a perfect phonograph reproducing the tones of the violin to the minutest detail, then he is entitled to assert that the curve described by the style on the disc of the phonograph provides, by its design and relief, an exact and complete description of the sonorous vibrations perceived by the listeners to the music. The extreme complexity of quality, attributed by the audience to the composition and the playing, are really properties of this curve, since the curve can reproduce it. This example shows clearly, as it appears to me, both the insufficiency and the great beauty of scientific description. How could such a description seem anything but extremely poor, as compared with the variety and wealth of our emotions and sentiments? It is easy to make merry at the expense of the scientist, when he professes to believe that he can find on the disc of the phonograph the genius of the composer or the emotion of the audience. On the other hand, the more formidable the disproportion between these infinitely varied human qualities and the apparent monotony of a collection of discs, the more reason we have to marvel that the latter can represent the former at all. In scientific language we can express this fact by saying that all acoustic phenomena can be completely described by means of a simple mechanical image, since we have only to set the disc under the style in motion in order to reproduce these phenomena. Moreover, it is as important simply to be aware that this mechanical reproduction is possible, as it is to understand the practical means by which we can realize it.

92. Mechanical Representations.

What do we understand exactly by a mechanical representation of a phenomenon? It is a representation the complete description of which requires only the numerical knowledge of data defining the motion of certain portions of matter, portions the forms of which can themselves be specified by numerical data. The definition of motion requires the measurement of space and time; the definition of a portion of matter requires the measurement of space, but it also requires a knowledge of the chemical nature of the substance. This nature may be very complicated, and thus quality in all its variability is introduced into our attempt to give a quantitative explanation.

It is a step in advance when we succeed in defining all chemical substances by means of a finite number of simple atomic elements. The maximum of simplicity would evidently be reached if all these atomic elements could be reduced to one single element, which we will call, for instance, hydrogen, and with which the most complicated molecular edifices would be built up. This hypothesis of the unity of matter has attracted many minds ever since it was found that the atomic weights of many bodies, considered to be simple, are very approximately exact multiples of the atomic weight of hydrogen. It would take too long to retrace here, even summarily, the history of this hypothesis, upon which the theory of isotopes [1] has recently shed a new light. We may provisionally admit this unity of matter; but we are not yet at the end of our difficulties.

There are in fact some very important physical phenomena, in heat, light, electricity, and magnetism, which appear to be of a nature entirely different from purely mechanical phenomena. It is many years since physicists succeeded in giving an explanation of thermal phenomena by considering heat as the perceptible manifestation of movements hidden to our senses, the molecules of hot bodies being supposed to be in a state of constant agitation, which is the more rapid the higher the temperature. Numerous experimental facts have convinced physicists of the reality of this thermal agitation, although it is not directly perceptible to our senses. With regard, however, to light, electricity, and magnetism, in spite of numerous efforts made by many scientists during the nineteenth century, we have not yet been able to establish a satisfactory mechanical theory, that is to say to give a complete description of the phenomena in terms of matter and motion. Maxwell, however, by his electromagnetic theory of light, has reduced the description of optical phenomena to that of electrical and magnetic phenomena, so that the complete description of the universe requires, in addition to matter and motion, a knowledge of the electromagnetic state at every point of space. The equations which govern mechanics, that is to

[1] See Jean Perrin, *Les Atomes*, 11th edition.

say, determine the movements of matter, are the famous equations established by Lagrange towards the end of the eighteenth century; the equations which govern electromagnetism are the not less famous equations of Maxwell.

Such was the state of science at the end of the nineteenth century. Physicists and mathematicians were endeavouring to establish a link between the equations of Lagrange and the equations of Maxwell, or, to use a phrase apparently more concrete, to define the relations existing between matter and ether, since it was agreed to give the name of ether to the mysterious support of electromagnetic phenomena. A whole book of the size of the present would not be too long to relate all the stages in the history of these efforts towards unification. Let us mention only, with regard to the theory of electrons, the names of Lorentz and Poincaré, and come to the denouement, that is to say, the theory of Einstein. This denouement is, of course, also a prologue, since there is no finality in science.

93. Einstein's Purely Geometrical Representation.

Every physical phenomenon takes place at a definite moment and at a definite place. We shall not discuss the difficulties involved in the measurement of time and space, although Einstein's solution of these difficulties is fundamental. We shall obtain a first description of the phenomena if we possess a complete knowledge of the totality of the numerical data which enable us to locate them in space and in time. But this description of the phenomena will, we can see, be incomplete, since it is not sufficient to know only where and when an event happens; we must also know *what* is happening; whether it is, for instance, a question of the displacement of a material body, of an electrical current, or of a solar ray. So far as a complicated mathematical theory can be expressed in ordinary language at all, we might sum up Einstein's theory by saying that a full and complete knowledge of space and time relations is sufficient for a description of the universe. The nature of phenomena, and in particular the localization of matter and electricity, are deduced by means of simple formulæ from these relations of space and time. We have thus attained the highest degree of simplicity which, as it seems, we have any right to hope for, since all the numerical data refer to measurements of the same nature; the quality of the phenomena is entirely reduced to their quantity. The measurements of space and the measurements of time may be considered as being of the same nature; we know, in fact, that an interval of time is measured by the space traversed by light during that interval. The unity thus established does not, of course, get over the fact that a detailed analytical description is not necessarily the most perfect description of phenomena. Two hundred thousand accounts such as those given by Fabrice[1] would not constitute the best story of the battle of Waterloo; and, similarly, a detailed account, supposing such a thing to be possible, of the history of the milliards of milliards of molecules in a grain of corn would not be a substitute for a good description of germination. By the side of the general theory there will always exist particular theories, probably becoming more and more numerous as time goes on, but the variety and complexity of these additions should not prevent us from recognizing the importance of the simple basic theory, the scientific interest of which is not less than the philosophical. The trees must not prevent us from seeing the wood, but neither should we ignore the fact that the wood consists of trees, the trees of cells, the cells of atoms, and the atoms of corpuscles in motion.

[1] The hero of Stendhal's *Chartreuse de Parme.*

94. The Gaps: Statistical Theories and Discontinuities: the Theory of Quanta.

In conclusion, I should like to say a few words with regard to the principal gaps which seem to me to exist in the profound theories of which we have tried to convey an idea, rough and superficial as this is bound to be. On the one hand, there is no solidly established link between these theories and the statistical theories, the essential importance of which I have shown elsewhere;[1] and, on the other hand, the question of the relations between the continuous and the discontinuous is again becoming one of living interest. It is for this reason that I have decided to reprint below (Note III) a few old reflections on a particular aspect of these relations. In reality it is the very bases of science that are being called in question, if it be true to say that modern science dates from the application of the differential calculus to the study of natural phenomena, that is to say from the use of the continuous variable. The history of mathematics, moreover, shows by numerous examples that the theories of the continuous and of the discontinuous have not ceased to react upon one another to their mutual advantage. It is certain that the differential calculus will not be thrown on the dust-heap, any more than Euclidean geometry and classical mechanics, and that they will remain the instruments best fitted to solve certain questions, questions which for us are of the highest importance, since they are related to our own scale of magnitude. It is, however, possible that an analysis and a geometry of the discontinuous will be necessary for the purpose of solving the problems raised by the physics of the discontinuous and at present summed up under the name of the quantum theory.

[1] See my book, *Le Hasard* (F. Alcan).

CHAPTER VIII

Recent Theoretical and Experimental Researches

95. The Equations of Electromagnetism.

The synthesis of the universe which the general theory of relativity professes to realize will only be complete when the equations include an account of electromagnetic phenomena also. It is a matter, in short, of explaining by a system of equations not only the phenomena relating to space, time, and matter, but also electrical and magnetic phenomena.

We cannot enter here into the details of the various attempts which have been made in this direction, notably by Weyl, Eddington, and Th. de Donder. Their main feature is the introduction, in addition to the potentials $g_{\mu\nu}$, of other functions $a^{\rho}_{\mu\nu}$ which are at once more numerous and of a more complicated nature. The number of these new potentials is forty, if we assume that they are symmetrical with respect to the lower indices μ and ν. There are in fact ten combinations two by two of these indices, and four possible values for the upper index. But these forty potentials $a^{\rho}_{\mu\nu}$ are not independent of one another, nor of the potentials $g_{\mu\nu}$. The relations connecting these diverse elements are necessarily very complicated, and give rise to almost inextricable calculations. We thus have here a field of studies which deserves to be explored, but of which only the first difficulties have been approached.

96. The New Mathematical Theories.

It is possible that these difficulties will be solved only after a preliminary advance in mathematical theories. We ought to remember, in fact, that if it was possible to develop the general theory of relativity so rapidly, this was due to the fact that nearly thirty years ago Ricci and Levi-Civita had created a mathematical apparatus exactly fitted for the development of this theory. The absolute differential calculus of Ricci and Levi-Civita,[1] thanks to a comparatively easy transformation and extension, has produced the tensorial calculus which is constantly being employed in the general theory of relativity.

It is a characteristic rôle of the mathematicians thus to create abstract theories which, at the moment when they are created, seem mere pretty ideas, but which afterwards turn out to be of value for immediate and urgent needs. The most celebrated instance of this is the theory of conic sections, which the Greek geometers had studied for their simple geometrical interest, but by the help of which Kepler was able twenty centuries later to formulate the laws of the motion of the planets. More recently Cauchy's theory of functions of a complex variable has been found useful in the calculations of technical electricity. It is now employed regularly, and has been a valuable aid in the theory of the propagation of waves, including the equations of Maxwell and Hertz, with which one must directly associate the invention of wireless telegraphy. The mathematical instrument necessary for the creation of the new physical theory will be found in an extension of the notion of space. The researches on the subject of the linear element made by Gauss, Riemann, Beltrami, Darboux, and many other geometers whom I cannot quote, have played a very important part. At present numerous geometers are endeavouring

[1] Cf. *The Absolute Differential Calculus*, by T. Levi-Civita. (Blackie & Son, Ltd., 1927.)

to generalize these results in various directions. We may mention the Italian school deriving its inspiration from the work of Levi-Civita, and in France the extremely profound and penetrating researches of Cartan, associated with the school of Darboux. Many years, however, will pass before these theories are complete.

97. Their Physical Significance still to be Found.

It appears to us also that such attempts at synthesis as those made by Th. de Donder, in spite of their analytical interest, in spite of the admiration one must bestow upon the effort which they represent, are still premature. Even Th. de Donder himself does not hesitate to admit that some of his equations are a little in the air, as it were, in the sense that it is impossible to assign to them any precise physical significance.

It seems, therefore, that for some time at least it would be preferable for mathematicians to apply themselves to the work of improving their instruments without devoting their attention unduly to physical theories. It should be sufficient for them to be more or less vaguely inspired by the wants of physicists to direct their mathematical researches in this direction or in that. When the mathematical apparatus is sufficiently perfect, then it will be the turn of the physicist to find out under what form the synthesis of general relativity and of electromagnetism can best be realized.

Whilst awaiting this result, it is desirable to follow very closely experiments designed to test the very foundations of the theory. Although these experiments, as we shall see, do not seem for the moment really to yield new and decisive conclusions, there is no reason why they should not be continued and followed with the greatest care.

98. Miller's Experiments.

A great stir has recently been caused by the experiments made in the United States by Professor D. C. Miller who, with a larger apparatus, repeated the famous experiment of Michelson and Morley.

As we have shown in § 66, this oft-repeated experiment has hitherto yielded negative results, that is to say results opposed to those which might be expected according to the old theory with regard to the ether. According to this theory, the motion of the earth in motionless ether will be indicated by a sort of ether drift, which will be revealed by a displacement of the interference fringes when the orientation of the apparatus is changed. Now, these fringes, far from being displaced by the expected amount, have shown only small variations which could be attributed to redundant influences (such as temperature, magnetic effects, &c.). It was experiments of this type which Professor Miller made in 1921, 1924, and 1925, and from which he drew conclusions which he communicated to the National Academy of Sciences of Washington, republished in the American Magazine, *Science*, of 19th June, 1925.

Professor Miller begins by giving a brief historical statement, recalling the fact that after having taken part (in collaboration with Professor Morley) in the very exact experiments made at Cleveland in 1904 and 1905, he and his collaborator had already concluded that positive results were possible at high altitudes. " We may therefore declare," they say, " that the experiment shows that if the ether near the apparatus did not move with it, the difference in velocity was less than 3·5 kilometres per second, unless the effect on the materials annulled the effect sought. Some have thought this experiment only proves that the ether in a certain basement room is carried along with it. We desire therefore to place the apparatus on a hill to see if an effect can be there detected."[1]

It was after the publication of the results of the experiments of 1904 and 1905 that Einstein intervened by publishing his first memoir on the theory of relativity. Professor

[1] *Science*, 19th June, 1925, p. 619.

Miller could never accept this theory. " This interpretation of the experiment," he says, " was not acceptable to the writer, and further observations were undertaken to determine this particular question."[1]

For this purpose an interferometer, analogous to that of 1905, that is to say permitting a light path of 224 feet was mounted on Mount Wilson observatory near Pasadena (California) at a height of about 600 feet. According to Professor Miller's account " 5000 single measures have been made at various times of the day and night. These have been reduced in 204 different sets, each consisting of observations made within 1 hour's time. The observations correspond to four different epochs of the year, as follows: 15th April, 1921, 117 sets of observations; 8th December, 1921, 42 sets; 5th September, 1924, 10 sets; and 1st April, 1925, 35 sets." They gave " a positive effect such as would be produced by a real ether drift corresponding to a relative motion of the earth and ether of about 10 kilometres per second."[2]

All sorts of precautions were taken in order to avoid perturbations of a magnetic or thermal nature; the original steel frame was completely covered with cork about one inch thick, then replaced by a frame of concrete reinforced with brass. " Many variations of incidental conditions were tried, observations were made with rotations of the interferometer clock-wise and counter clock-wise, with a rapid rotation and a very slow rotation. . . . Various sources of light were employed, including sunlight and the electric arc. . . . The results of the observations were not affected by any of these changes."[3]

Professor Miller concludes his communiqué in the following terms: " The ether experiments at Mount Wilson during the last four years, 1921 to 1925, lead to the conclusion that there is a positive displacement of the interference fringes, such as would be produced by a relative motion

[1] *Science*, 19th June, 1925, p. 619. [2] *Ibid.*, p. 619. [3] *Ibid.*, p. 620.

of the earth and the ether at this observatory of approximately 10 kilometres per second, being about one-third of the orbital velocity of the earth. By comparison with the earlier Cleveland observations, this suggests a partial drag of the ether by the earth, which decreases with altitude. It is believed that a reconsideration of the Cleveland observations, from this point of view, will show that they are in accordance with this presumption and will lead to the conclusion that the Michelson-Morley experiment does not give, and probably never has given, a true zero result.[1] He adds that in his opinion one should be able to discover in this manner not only the motion of the earth round the sun, but also the absolute motion of the latter. He concludes his article with the remark that a complete calculation, to be made in the immediate future, should give definite indications regarding the absolute motion of the solar system in space.

At the moment of going to press none of the numerical details promised by Professor Miller have reached us. Now it is just these numerical details which are indispensable for a serious discussion of the experiment. We must also point out that in his article Professor Miller gives only "conclusions", "suggestions", and "interpretations", and his interpretations are those of a man who, as he admits himself, has never considered Einstein's theory as acceptable. He who listens to only one bell, hears only one sound.

99. Miller's Experiments and other Phenomena.

But the interpretation suggested by Professor Miller clashes with a fact of observation whose laws have been known these many years to a high degree of accuracy—we mean the phenomenon of the aberration of the stars. This phenomenon (described above in § 67) shows that the luminous waves are not carried along by the earth, whether at the bottom of the valleys or upon the mountain summits. And the conclusions of the kinematic argument,

[1] *Science*, 19th June, 1925, p. 621.

starting from the hypothesis that there is no dragging of the waves, are at least verified by facts, whilst the deductions drawn from the hypothesis of drag are not.[1] As for the phenomenon of partial drag of the waves in refracting surroundings in motion, the theory of which was given by Fresnel and verified by Fizeau (see above, § 69), this phenomenon, too, is at variance with the theory of the drag of the ether by the earth.

All these phenomena, however, are, as we have shown above, completely explained by the theory of relativity. " Evidently," writes Einstein himself in a letter addressed to the magazine *Science*, " evidently, if Dr. Miller's results should be confirmed, then the special relativity theory, and with it the general theory *in its present form*, fails. Experiment is the supreme judge." " But," adds Einstein, " we must take into account the fact that no theory exists outside the theory of relativity and the similar Lorentz theory which, except for the Miller experiment, explains all the known phenomena up to date." And he concludes: " Under the circumstances, nothing remains but to await a more complete publication of Miller's results. Then it is to be hoped that a correct decision will develop."

Indeed, Miller's results (in spite of the scantiness of the details so far published) seem already to need to be attributed to causes which are quite different from the famous " ether drift ". Professor Miller himself says in fact that, in addition to the deviation of the fringes repeating itself at every half-turn of the interferometer, he had observed another important deviation superposed upon the first, with a period corresponding to a complete turn of the apparatus. He endeavours to explain this in the following words: " An extended investigation in the laboratory demonstrated that the full period effect mentioned in the preliminary report

[1] See on this subject the notice of A. Metz, communicated to the Académie des Sciences of Paris on 9th February, 1925 (*C. R. de l'Acad. des Sc.*, t. 180, 1925, p. 495).

on the Mount Wilson observations is a necessary geometrical result of the adjustment of mirrors when fringes of finite width are used.[1] But he gives no demonstration of the geometrical result asserted to be necessary. In reality it is always on the cards that the explanation (still to be found) of the full-period effect will also furnish the key to the half-period effect attributed by Miller to the ether drift. As M. Brylinski[2] has suggested in a note communicated to the Académie des Sciences of Paris, it is quite possible that the perturbations observed may be of a thermal nature; a slight inequality in the lighting of the room where the interferometer is installed would be sufficient to explain periodical effects analogous to those observed. Professor Miller has anticipated this objection, and replies that " four precision thermometers were hung on the outside of the house. On several occasions the extreme variation of temperature was not more than 0·1 degree and usually it was less than 0·4 degree." He asserts that " such variations did not at all affect the periodic displacement of the interference fringes ", which is quite inexact, as the extension corresponding to 0·1 degree is of the order of $1/10^6$ for the ordinary metals and for concrete, whilst the phenomenon to be observed corresponds to a variation of the length of the arm of the apparatus of the order of $1/10^8$.

100. Michelson and Gale's Experiment.

Moreover, an experiment has recently been made by Michelson himself in collaboration with Gale, also professor at the university of Chicago, and the results are directly opposed to the interpretation given by Professor Miller. This experiment, entirely different from the famous experiment of Michelson-Morley, was to enable the authors to test the effect of the earth's rotation on the velocity of light (whilst the experiment of Michelson-Morley should

[1] *Science*, 19th June, 1925, p. 620.
[2] *C. R. Acad. des Sc.*, Paris (t. 79, 1924, p. 559), 22nd Sept., 1924.

have revealed the effect of the translation of the earth through the ether). A report on the experiment was given in the scientific journal, *Nature*.[1] The experiment was almost analogous to one made in Paris by Professor Sagnac a few years before for the purpose of putting in evidence the rotation of an interference apparatus by means of an optical effect. This apparatus was composed of a frame furnished with mirrors which caused a ray of light to make a complete turn in one sense, whilst another ray from the same source was sent round in the opposite direction so as to interfere with the first. When the whole apparatus was given a rotary motion the interference fringes were displaced. This displacement was found to be equal to that obtained by calculation on the basis of the classical theory of the ether. But Einstein's theory, too, accounted perfectly for the result, as has been shown by Professor Langevin, both by the special theory of relativity and by the general theory.[2]

In short, the experiment of Michelson and Gale is the same as that of Sagnac, but carried out with apparatus on a much larger scale, fixed on the earth in such a manner as to show the rotation of the earth.

The result is perfectly well explained by the theory of relativity, as was shown by Professor Larmor and Dr. Jeans in a later number of *Nature*.[3]

The latter has clearly shown the consequences of the Michelson-Gale experiment: "Freed of all hypotheses about the ether," says Jeans, "the experiments appear to show that the velocity of light in space is the same (to within one part in 10^{11}), whether the light travels in the direction of the earth's rotation or in the contrary direction. This is in accordance with the theory of relativity. Thus the experiments do not affect the position of this theory,

[1] *Nature*, 18th April, 1925, p. 566.
[2] *Comptes rendus Acad. Sc.*, 7th Nov., 1921.
[3] *Nature*, 2nd May, 1925, p. 638.

although a contrary result would have destroyed the theory. The experiments show either that there is no ether or else that—if there is an ether—the earth does not drag this ether into motion by its rotation.

" The original Michelson-Morley experiment admits of three separate interpretations:

" (*a*) There is no ether.

" (*b*) There is an ether which accompanies the earth in its motion.

" (*c*) There is an ether which is at rest in space, bodies moving through it undergoing contraction in accordance with the Lorentz-Fitzgerald formula.

" The present experiments dispose of interpretation (*b*), which, however, is generally supposed to be adequately disposed of already by the phenomenon of astronomical aberration."

This interpretation is just that suggested by Professor Miller as the result of his experiments on Mount Wilson (he adds that the ether is less dragged upon the mountains than on the plains). The argument from the aberration of light, for the discussion of which some scientists have proposed to postulate certain deformations of the ether,[1] is thus strengthened by a " purely optical experiment, free from need of confirmation by other determinations, whether directly parallactic or indirect results of gravitational astronomy ", as Sir Joseph Larmor says.

Thus the Michelson-Gale experiment, the results of which have been published *in extenso* and could therefore be fully and openly discussed, far from being contrary to the special theory of relativity, have rather confirmed it by eliminating Professor Miller's interpretation of the " partial drag of ether by the earth ".

[1] See on this subject the notice of M. Ferner, presented to the Acad. des Sciences of Paris at the meeting of 22nd June, 1925 (*Comptes rendus*, Vol. 180, p. 495).

It has been the same up to now with all experiments which have been brought forward as contrary to Einstein's theories. It will probably be the same with Miller's experiments also when the details are finally made public, and a serious examination of the results by the scientific world becomes possible.

101. The Detractors of the Theory of Relativity.

We must not expect, however, the systematic detractors of the theory of relativity to disarm at once. A curious consequence of the celebrity so rapidly attained by this theory was the attention bestowed upon it by a number of people who are not usually in the habit of taking an interest in mathematical and physical speculations. Popular articles published in reviews and even in daily newspapers have given these people a frequently wrong and always incomplete idea of the theory of relativity. Moreover, the efforts, often laudable, made by the popularizers to render the theory comprehensive have had the result of persuading many people that they have mastered all the details of the theory, whilst in reality they have at the utmost acquired only a general outline.

Hence arose the idea that a theory which could be explained in the few pages of a review article or in a column of a daily paper could also be critized or refuted in an article of the same length.

Those who have undertaken the task of refuting the theory do not realize that behind the review articles, simple résumés of the theory, there exist a vast number of studies, published either in special journals or as separate works, and that those innumerable purely technical studies constitute a coherent mass which it is very difficult to shake or unsettle. It is not by discussing the terms of a review article, or even of a small work like the present, that one can hope to demolish a theory the essentials of which can only be explained with the help of numerous developments and innumerable mathematical formulæ.

Those who are anxious to refute or even to discuss the theory of relativity should first take the trouble to study it thoroughly, but only very few people have taken this trouble. Mathematicians and physicists whose previous training and studies put them in a position to make such a deep and thorough investigation and who had at first been somewhat sceptical with regard to the new theories, have now adopted a more prudent attitude and have ceased to write on the subject of the theory of relativity.

We must therefore warn the public against these ostensibly scientific publications which are only the simple results of the author's imagination and are in no way based upon any serious foundation, failing as they do to take account of the vast number of studies on the theory published during the last few years.

102. The Misconceptions of the Philosophers.

In addition to these apparently scientific refutations of the theory of relativity, we must also mention the very numerous philosophical studies to which the theory of relativity has given rise or for which it has served as a pretext.

It would appear that the majority of these philosophical studies are based on a sort of play upon words. The term "relativity" has for a long time been employed by the philosophers with a more or less precise meaning. Speaking generally, the word *relative* in philosophical language is opposed to *absolute*. The term *relativity* is used by Einstein and the physicists in an undoubtedly legimate but nevertheless special sense.

Scientific language is necessarily much more precise than philosophical and popular language, so that scientists are compelled either to invent entirely new words or to borrow words from ordinary language, defining the precise sense in which they are to be used. It is thus that the term *function* has for mathematicians a very narrow sense, differing from the various senses it possesses in ordinary language. Such is also the case with regard to the word *relativity*, which one should avoid employing except with the addition of the epithets *special* or *general*. These expressions denote physical and mathematical theories, and it is really an abuse of language to endeavour to interpret them while sticking to the ordinary meaning of the term relativity.

Apart from this, we see nothing to regret in the fact that the popularity of Einstein's theory has led the philosophical public to reflect upon relativity in general, and it is quite possible that this reflection has resulted in interesting results. This, however, is a simple verbal coincidence, and these results are entirely outside our subject.

103. It is now the Turn of Experiment.

If one conclusion rather than another is to be drawn from the present work it is that of the experimental character of the new conception of space and time to which the theory of relativity leads. We are very willing to believe that we have the intuition of some sort of universal time, and it is the existence of this feeling of intuition which is the cause of the instinctive repulsion experienced by those who are faced for the first time by the new theories. A little reflection, however, will be sufficient to make them understand that if we talk, for example, of *the event which is taking place upon the sun at the moment whilst I am speaking*, the precise sense of the expression can be defined only by appealing to certain experiments, either real or ideal. Both the problem of time and the problem of space connected with it are thus transferred from the philosophical and theoretical to the experimental domain. This is a definite philosophical advance which it will be impossible ever to lose. But we must not lose sight of the fact that the very adoption of this point of view involves, so far as the adherents of the new theory are concerned, the strict obligation to turn to experiment as sovereign judge. They should furnish experimental proofs of the new theory. We have already indicated a certain number of such proofs, but are bound to admit that relatively their number is still limited, and that, on account of the precision they require, they are of an extremely delicate nature.

We cannot, of course, hope to be able to demonstrate the theory of relativity by means of phenomena on a gross scale, but we are entitled to think that the developments of experimental technique will enable us to predict and to verify a greater number of phenomena accessible to all careful experimentalists. On the day on which this desideratum is realized, there will be no detractors of the theory of relativity, as there are none of Newton's cosmology.

104. **Supplementary Note.**

During the passage of this edition through the press, one or two works have appeared which deserve to be noticed. From the experimental point of view, M. Esclangon, director of Strasbourg Observatory, has tried to demonstrate by astronomical methods the translational motion of the solar system. The results he has obtained, which have appeared in the *Comptes Rendus* of the Academy of Sciences (first half-year, 1926), require confirmation by new observations. If they are confirmed, the question will arise whether they cannot be explained, in the general theory of relativity, by the existence of a *curvature*.

On the other hand, M. Eyraud has sustained, in June, 1926, before the Faculty of Sciences of Paris, a thesis on " Metric Spaces and Physico-Geometric Theories ". Making use of the conception of general torsion due to M. Cartan, he succeeds in developing a mathematical theory of the relations between gravitation and electromagnetism which appears to possess real advantages over the theory of M. H. Weyl.

NOTE I

The Kinematics of the Special Theory of Relativity

As we have seen, Michelson's experiment leads us to conclude that it is impossible by optical or electromagnetic means to detect motion relative to the ether. We must infer from this that the equations of electromagnetism define a continuous group of transformations, with respect to which they are invariant, and that this group, which we call the Lorentz group, defines the space and time relations between two systems in relative translational motion, the axes used in the two systems having their directions fixed with respect to the stars. The invariance of the equations of electromagnetism demands the invariance of D'Alembert's function, by means of which the propagation of waves is defined:

$$\frac{1}{c^2}\frac{\partial^2\phi}{\partial t^2} - \Delta\phi, \ldots\ldots\ldots\ldots\ldots (1)$$

in which c is the velocity of light and $\Delta\phi$ the function of Laplace:

$$\Delta\phi = \frac{\partial^2\phi}{\partial x^2} + \frac{\partial^2\phi}{\partial y^2} + \frac{\partial^2\phi}{\partial z^2}, \ldots\ldots\ldots (2)$$

which is invariant under orthogonal transformations in ordinary space. The invariance of the Dalembertian is equivalent to that of the quadratic form:

$$ds^2 = c^2dt^2 - dx^2 - dy^2 - dz^2, \ldots\ldots (3)$$

by which we have defined the elementary interval. The transformations which leave the form (3) invariable can be obtained by combining those which involve x, y, z only and those which involve x and t. We shall confine ourselves to these last; we can write them:

$$\left.\begin{aligned} X &= \frac{x + \lambda ct}{\sqrt{(1 - \lambda^2)}} \\ cT &= \frac{\lambda x + ct}{\sqrt{(1 - \lambda^2)}} \\ Y &= y \\ Z &= z \end{aligned}\right\} \qquad (4)$$

We see that if we put $X = 0$, we get $x = -\lambda ct$; the product λc is, therefore, equal in absolute value to the velocity v_0 of the origin of the moving axes with respect to the fixed axes. To obtain the formula for the composition of velocities, it is sufficient to calculate $\frac{dX}{dT}$ in terms of $\frac{dx}{dt}$; we find

$$\frac{dX}{dT} = \frac{\frac{dx}{dt} + \lambda c}{1 + \frac{\lambda c}{c^2}\frac{dx}{dt}}, \qquad (5)$$

or, putting

$$\lambda c = -v_0, \quad \frac{dx}{dt} = v_2, \quad \frac{dX}{dT} = v_1,$$

$$v_1 = \frac{v_2 - v_0}{1 - \frac{v_0 v_2}{c^2}}, \qquad (6)$$

or, what comes to the same thing,

$$v_2 = \frac{v_1 + v_0}{1 + \frac{v_0 v_1}{c^2}} \qquad (7)$$

The formula (7) is the formula of composition of velocities in the theory of relativity; v_0 is the velocity of the system X, T with respect to the system x, t; v_1 is the velocity of a point connected to the system X, T, estimated with reference to this system; v_2 is the velocity of the same point with reference to the system x, t (what is sometimes called its absolute velocity). We see that if the velocity c of light were replaced by infinity the formula (7) would coincide with the formula of classical mechanics, $v_2 = v_0 + v_1$. If $v_0 = c$, the formula (7) gives also $v_2 = c$; we thus find, as we might expect, that the velocity of light is invariant—a consequence of the invariance of (1), where c plays the part of the velocity of propagation of the disturbance.

Fizeau's experiment on the drag effect of running liquid on waves of light is equivalent to taking $v_1 = \frac{c}{n}$, for the velocity of light in a liquid is equal to the quotient of the velocity *in vacuo* (or in air) by the index of refraction of the liquid; the formula (7) therefore gives:

$$v_2 = \frac{v_0 + \frac{c}{n}}{1 + \frac{v_0}{nc}} = \frac{c}{n} + v_0\left(1 - \frac{1}{n^2}\right) + \ldots, \quad \ldots(8)$$

the terms omitted containing $\frac{1}{c}$ as a factor; this represents the partial drag effect demonstrated by Fizeau, obtained by applying the laws of classical kinematics to a running liquid whose velocity is $v_0\left(1 - \frac{1}{n^2}\right)$ instead of v_0.

The ease with which this confirmation of a classical experiment follows by elementary calculations, beginning with the Dalembertian (1), which plays a classical part in optics, appears to me to be one of the most striking pieces of evidence in favour of the theory of relativity.

The formulæ (4), when solved for x, y, z, and t, give:

$$\left.\begin{aligned} x &= \frac{X - \lambda cT}{\sqrt{(1 - \lambda^2)}} \\ ct &= \frac{-\lambda X + cT}{\sqrt{(1 - \lambda^2)}} \\ y &= Y \\ z &= Z \end{aligned}\right\} \quad \cdots\cdots\cdots\cdots (9)$$

The formulæ (9) only differ from (4) in the change of λ into $-\lambda$, that is to say, in the sign of the relative velocity λc.

If we put $t = 0$ in (4) we obtain:

$$X = \frac{x}{\sqrt{(1 - \lambda^2)}}; \quad \cdots\cdots\cdots\cdots (4')$$

and if we put $T = 0$ in (9) we obtain:

$$x = \frac{X}{\sqrt{(1 - \lambda^2)}}. \quad \cdots\cdots\cdots\cdots (9')$$

The results (4′) and (9′) would be inconsistent if t and T were identical, which they are not. The two results together are the expression of the *Lorentz contraction*, in other words of the fact that a moving length appears to be shortened in the direction of motion, for observers with respect to whom it is in motion. The factor of reduction is $\sqrt{(1 - \lambda^2)}$, the value of λ being $\frac{v}{c}$, the ratio of the relative velocity to the velocity of light. In the case of formula (4′) we have taken $t = 0$, so that we have supposed ourselves part of the system x, t, considered as at rest; the distance between two points X_1, X_2, when measured with respect to this system, is $x_1 - x_2 = (X_1 - X_2)\sqrt{(1 - \lambda^2)}$. On the contrary, in the formula (9′) we have taken $T = 0$, that is we have taken the system X, T as the one at rest, and the new formula for the lengths is inverse to the other.

Another consequence of the kinematics of relativity is the change of apparent mass (or inertia) with velocity. This is the dynamical equivalent of the purely kinematical conception of a velocity c which can never be reached: there could be no such velocity unless the speed of a mass became more and more difficult to increase as it approached the value c, i.e. unless mass increased with speed. The increase of inertia with velocity has been verified experimentally by Kaufmann in his experiments, now classical, on the β-particles of radium (1902–6). These experiments have led to a large number of laboratory researches and theoretical discussions, which it would be impossible to summarize unless by entering upon complete explanations of the dynamics of relativity. Suffice it to say that two main theories have been propounded, that of Lorentz and Einstein, based upon the kinematics and dynamics of relativity, and that of Max Abraham, based on Newtonian mechanics. M. Ch. Eug. Guye, the learned professor of physics at the University of Geneva, has published (Dec., 1921) detailed results of experiments which he has carried on for many years, in collaboration with S. Ratnowsky and Ch. Lavanchy. His memoir, entitled *Vérification expérimentale de la formule de Lorentz-Einstein*, occupies nearly 100 pages in the memoirs of the Geneva Society of Physics and Natural History, with numerous numerical tables. The experimental results agree perfectly with the Lorentz-Einstein theory, showing only trifling deviations, both positive and negative, thoroughly in keeping with the regular character of experimental errors; on the contrary, with Abraham's theory, the deviations are nearly all of the same sign, and the mean of their algebraical values is much larger (0·0112 instead of 0·0002). These very careful experiments therefore constitute a most striking confirmation of the theory of relativity.

NOTE II

On the Fundamental Hypotheses of Physics and of Geometry

Recent works on the theory of relativity, notably those of Weyl[1] and Eddington[2], have shown the importance for physicists of researches on the principles of geometry, researches sometimes regarded as relevant to philosophy rather to science. It may not, then, be useless to try to define what exactly are the fundamental hypotheses, of a physical character, which are implied in every theory which has aimed at a synthesis of geometry and physics. The essential ideas are due to Riemann and Poincaré; we only require to adapt them to the suggestions of experiment as interpreted by a Lorentz and an Einstein, suggestions which have caused us to regard as possible and even probable physico-geometrical facts which in the nineteenth century would have appeared most unlikely.

To simplify the language and the notation, I shall speak simply of three-dimensional space considered at a given instant; in other words, I shall discuss a section (not a narrowly restricted one) of the four-dimensional universe; analogous considerations can be applied to the universe itself.

The *first hypothesis* consists in the assumption that it is possible to define in space a network of co-ordinates in three dimensions, u, v, w. From the local point of view, this hypothesis is essentially a hypothesis of continuity; it is not, in fact, possible to suppose actually traced a continuous infinity of surfaces $u =$ constant; it is therefore

[1] H. Weyl, *Raum, Zeit, Materie*. French and English translations.

[2] A. S. Eddington, *Space, Time, and Gravitation* and *Mathematical Theory of Relativity*; also *Proceedings of the Royal Society of London* (A, Vol. 99, May, 1921).

necessary to admit the possibility of interpolation, subject to the single condition that a sufficiently great number of these surfaces has been chosen and numbered; this is the hypothesis of continuity. From the general point of view, the form of the network is related to that of space as a whole considered from the standpoint of *analysis situs*; and any hypothesis which is not logically absurd is compatible with the facts observed in the portion of space accessible to our investigations, provided only the hypothesis relates to portions of space far enough away. It appears superfluous to introduce a hypothesis which we cannot verify.[1]

The network u, v, w being supposed defined, consistently with the hypothesis of continuity, imagine that at each point of space we have an infinitely small measuring-rod or gauge, and that we wish to find the locus of one of its ends, the other remaining fixed. The *second hypothesis* is that this locus is an ellipsoid which has its centre at the origin, in the system of co-ordinates u, v, w; this hypothesis, which we shall assume to be verified whatever may be the physical nature of the gauge and the physical state of the point of space considered, comprehends in itself numerous physical hypotheses or, if we prefer to say so, the results of numerous experiments. This second hypothesis allows us to write the value of ds^2 for the system of co-ordinates adopted, in the form

$$ds^2 = \lambda^2\phi(du, dv, dw), \quad \ldots\ldots\ldots\ldots(1)$$

where ϕ is a quadratic form with variable coefficients, corresponding to the ellipsoid defined at each point, and λ is an arbitrary factor characterizing the variable length of the gauge at different points of space. If we assume that we take for the gauge at different points the " same " physical magnitude, there will follow the continuity of λ

[1] I do not consider the number *three* for the dimensions of space to be a hypothesis, but a physical fact.

and of the coefficients of ϕ; in any case, the coefficients of the form $\lambda^2\phi$ must be supposed to be in general continuous; this is the *third hypothesis*.[1]

It is permissible, if it suits us, to add to these three physical hypotheses a *fourth hypothesis* which we may call *geometrical*, to a large extent arbitrary, which will consist in defining a geometrical $d\sigma^2$ by the formula

$$d\sigma^2 = \psi(du, dv, dw), \quad \ldots\ldots\ldots\ldots(2)$$

where ψ is any quadratic form (with continuous coefficients); if the curvature belonging to this form is null, the geometrical space will be Euclidean.[2]

The relation between the geometrical space (2) and the physical space (1) is given by the formula

$$\frac{ds^2}{d\sigma^2} = \frac{\lambda^2\phi(du, dv, dw)}{\psi(du, dv, dw)},$$

in which the invariants of the second member, with reference to general point transformations, express the physical properties of the geometrical space we are considering, with respect to the gauges used (the geometrical space depends, moreover, on the particular choice of the section of the universe; this difficulty would not present itself if we had chosen four variables instead of three).

The introduction of the arbitrary function ψ may seem to be a useless complication; and it undoubtedly would be so if we chose it at random, just as it would lead to troublesome complication in classical mechanics to treat axes rigidly connected to the earth or to a definite atom as if they were absolute axes.

Only experience will enable us to decide whether certain

[1] This hypothesis is necessary on account of the impossibility not merely of carrying out, but even of imagining, the experiment with the gauge at *every* point of space; it is necessary to assume that a finite number of experiments will enable us to predict the results of all such experiments.

[2] Unless we take up a purely local standpoint the two forms ϕ and ψ must be compatible from the point of view of *analysis situs*.

choices of the function ψ allow us or not to give a simpler form to the physical properties of the universe; it will then perhaps be possible to attribute to one particular geometry (and one particular definition of time) a quality of truth based on the idea of convenience, in the sense of this word as used by Poincaré with reference to the rotation of the earth round the sun. Meantime, we should reserve the answer to the question whether the introduction of ψ is a superfluous complication or an almost indispensable simplification.

But, before taking up the fourth hypothesis, the proper procedure would be to make a close examination of the three first, and to inquire especially whether they continue to be well-founded on the molecular scale, or express truths which are merely statistical.

NOTE III

The Mathematical Continuum and the Physical Continuum[1]

1. The Scale of Rational Numbers.—I need not recall here the meaning of *whole number*, or of *rational number*. Nor do I discuss the psychological or metaphysical origin of these conceptions, for mathematicians are in such complete agreement when they speak of whole or of rational numbers that we can regard their accord in this respect as an established fact, on which we can build as upon solid ground.

Let us recall the chief properties of the aggregate of rational numbers.

[1] This note reproduces without modification an article published in 1909 in the *Rivista di Scienza*; I mention the date specially, because I discuss a theory of Poincaré's more freely than I could have done after the death of the illustrious mathematician, which occurred three years later (1912).

(1) This aggregate is *everywhere dense*. In other words, if we represent each number by that point on an axis whose abscissa is equal to the number, there are rational numbers on every piece of the axis, however small. This implies the following important consequence; the abscissa of any point whatever can, to as close an approximation as we wish—that is to say, with an error as small as we please—be represented by a rational number.

(2) Further, this aggregate is *countable* (enumerable). This means that we can arrange all rational numbers in a linear sequence, in such a way that each of them occupies a perfectly definite place; we can, for example, agree to write first those positive fractions for which the sum of the numerator and denominator has the smallest value; when this sum is the same for two fractions, we write first the one whose denominator is smallest; lastly, immediately after each positive number we write the corresponding negative number; we obtain in this way the following sequence:

$$0;\ \frac{1}{1},\ -\frac{1}{1};\ \frac{2}{1},\ -\frac{2}{1},\ \frac{1}{2},\ -\frac{1}{2};\ \frac{3}{1},\ -\frac{3}{1},\ \frac{1}{3},\ -\frac{1}{3};\ \frac{4}{1},\ -\frac{4}{1},\ \frac{3}{2},\ -\frac{3}{2},\ \frac{2}{3},\ -\frac{2}{3},\ \frac{1}{4},\ -\frac{1}{4};\ \frac{5}{1},\ \ldots$$

the law of formation of which is obvious; the only thing is that we must avoid writing down any fraction of the same value as one that has already occurred.

In general, we regard a rational number as the simpler the lower its order (the nearer the beginning it occurs) in the above or in a similar sequence. This law of simplicity becomes precise when we are given the law of formation of the sequence. It is often useless, however, to have this absolute precision; the more or less vague ideas of common sense will do; everyone would agree that the numbers 5 and $\frac{2}{3}$ are simpler than the numbers 2417 and $\frac{43}{7}$; but it may be worth while to observe that this notion of simplicity is essentially related to the fact that the aggregate of rational numbers is countable.

2. The Measurement of Magnitudes.—The measurement of magnitudes was the first application of the notion of number, and it is still the most important. To measure a magnitude is to express by means of a number, as exactly and as simply as possible, the ratio of the magnitude to the one chosen as unit. The two conditions of *exactness* and *simplicity* are sometimes difficult to reconcile; in that event we sacrifice the one or the other according to the practical conditions of the case. For example, if as the result of a series of measurements and calculations we obtain for the ratio of the diameters of two wheels

$$\frac{2}{3} + \frac{1}{1745287} = \frac{3490577}{5235861},$$

we shall probably prefer to adopt the simple fraction $\frac{2}{3}$ rather than the complicated fraction $\frac{3490577}{5235861}$, even if we have reason to think that the latter is more exact.

The theory of continued fractions allows us to point out a routine method for obtaining the simplest fractions approximating as closely as may be to a measured ratio. We often prefer, however, to use decimal fractions, on account of our familiarity with the rules of calculation on decimal numbers. This familiarity leads us to regard the fraction $\frac{17}{100}$, for example, as simpler than the fraction $\frac{14}{81}$, even though the terms of the latter are smaller. It is well to observe that the estimate of simplicity is here purely subjective; in the final analysis it is chiefly due to the fact that we have all learnt the multiplication table in the decimal system, that is to say that equalities such as

$$7 \times 8 = 5 \times 10 + 6$$

come up in our memories instantaneously.

Further, when we have to compare several fractions as to magnitude, it is convenient that they should have the same denominator, a condition easily realized if they are in the decimal form. Also, we find it easier to make an instinctive estimate of relative magnitude when we make use of the decimal scale; which explains why we sometimes prefer to say 33 per cent rather than *the third*, which is simpler and perhaps more exact. If, however, we are aiming at greater precision, we generally prefer to say or write $\frac{1}{3}$ of a metre, rather than 333·33 millimetres. This apparently trivial remark gives the key to the relations between the mathematical continuum and the physical, or again between the theoretical continuum and the practical.

The question arises, then, whether the exclusive use of decimal fractions may not be sufficient for practical requirements; in fact, when we give a workman the dimensions of an article he is to make, we use the decimal system, taking care to limit the number of decimals according to the precision attainable. Will it not be simpler, then, to confine ourselves to the consideration of terminating decimals? The aggregate of these possesses evidently the two essential properties of the aggregate of rational numbers—it is everywhere dense, and it is countable; the one aggregate can therefore render the same practical services as the other. Here is the sort of objection which we might raise to the use of other than decimal numbers: if we wish the width of a door to be $\frac{2}{3}$ of a metre, we tell the carpenter to make it 666 millimetres and 7 tenths, and the practical result wanted will be obtained with the utmost precision.

This objection contains a part of truth; it is perfectly correct to say that the use of decimal numbers carries with it practical advantages so great that we ought to be in favoûr of the abandonment of the relics of other systems of numeration which still survive, notably in the division of the circle, and in the division of time. Even if we agree to this, however, the evident fact remains that the use of decimals does not allow us to give certain relations the simplest form of which they are susceptible, *even if we take our stand on a purely practical ground.*

Consider for example this theorem of elementary geometry: *the medians of a triangle cut each other at a point which is a point of trisection of each of them.* This statement, like every mathematical statement, expresses two truths: (1) a purely logical truth, which we can obtain explicitly by replacing each term by its definition and starting from the axioms and postulates; (2) a practical truth, the only one with which I am concerned at present. This practical truth can be stated, or verified, in several ways; here are two typical examples. Consider a triangle drawn with care on a plane; let AM be a median, G the point between A and M where the medians cross each other. If we have measured the length AM and found it to be 475·9 millimetres, we infer that the length MG is equal to the product of this number by 0·3333, that is to say, is equal to 3 tenths of it, plus 3 hundredths of it, plus 3 thousandths of it, plus 3 ten-thousandths of it. That is the first statement—here is the second: if we mark off, end to end, with a pair of compasses for example, on the line AM, three lengths equal to MG, we cover exactly the length AM. The second statement is manifestly more simple and at the same time easier to verify than the first. In this sense, we can say that of the two following propositions:

(1) the length MG is equal to 3333 ten-thousandths of AM,
(2) the length MG is the third part of AM,

the second is not only the *only* one true in theory, but is also the *truer* of the two in practice. And what I mean by this is not that greater precision of measurement would prove the inexactness of the first proposition; because, if instead of the fraction 0·3333 we took the fraction 0·3333333333333333, we might be very certain that experiment would never disprove the assertion that this is the ratio of MG to AM. If it is right, from the practical point of view, to prefer the second statement, it is precisely because it is the simpler, and that in a twofold aspect: simplicity of form in the terms which express it, and essential simplicity, i.e. more immediate adaptation to effective use.

3. **Irrational Numbers.**—The Greek geometers were aware of the theoretical impossibility of measuring the diagonal of a square exactly, supposing the side of the square taken as the unit; the series of operations known as Euclid's algorithm continues indefinitely, leaving no loophole for the hope that, if we work on, it will stop sometime, for we can easily prove that it is periodic.[1] We are thus led to say that the ratio of the diagonal of a square to its side is the irrational number $\sqrt{2}$. What is the practical meaning of this statement? Why should we prefer it to the following: *the ratio of the diagonal to the side of the square is* 1·4142136? The two statements are equivalent if we merely take the point of view of immediate physical verification; the difference between $\sqrt{2}$ and 1·4142136 cannot be detected by measurement. It is not, therefore, because the value 1·4142136 is theoretically inexact that we ought *in practice* to prefer the value $\sqrt{2}$; for in *practice* the one value is just as exact as the other; but the value $\sqrt{2}$ is at once

[1] As we know, Euclid's algorithm is identical with the series of elementary operations by which we find the greatest common measure of two numbers; we divide the greater by the smaller, then the latter by the remainder from the first division, then this remainder by the remainder from the second division, and so on, until we arrive at a remainder which is nil; this always happens when we start from two whole numbers, but not always when we start from two geometrical magnitudes; when it does not happen the magnitudes are said to be incommensurable with each other. In the case of the diagonal and side of a square, we can easily prove by geometry that the ratio of the first remainder to the second is equal to the ratio of the second to the third, and so on; it is therefore impossible that the operation should come to an end. This geometrical fact is equivalent to the numerical identity

$$\frac{1}{\sqrt{2}-1} = \sqrt{2}+1 = 2+(\sqrt{2}-1),$$

which allows us to obtain the development as a periodic continued fraction

$$\sqrt{2}-1 = \cfrac{1}{2+\cfrac{1}{2+\cfrac{1}{2+\dots}}}$$

easier to remember and more convenient to handle in many calculations; numerical relations such as

$$(\sqrt{2} + 1)(\sqrt{2} - 1) = 1$$
$$(\sqrt{2} + 1)^3 = 7 + 5\sqrt{2}$$

are verified at a glance: the calculation of the expressions $(\sqrt{2} + 1)(\sqrt{2} - 1)$ or $(\sqrt{2} + 1)^3$ would on the contrary be very complicated if we were to replace $\sqrt{2}$ by 1·4142136; it would be necessary to carry out the following calculations:

$$2{\cdot}4142136 \times 0{\cdot}4142136,$$
$$2{\cdot}4142136 \times 2{\cdot}4142136 \times 2{\cdot}4142136,$$

calculations at once long and liable to error. Supposing for a moment that geometry had remained a purely empirical science, what would we know of the ratio of the diagonal of a square to the side? Measurements sufficiently exact would give for the value of this ratio 1·414; researches of high precision would permit us to state that it lies between 1·414213 and 1·414214. This will be the experimental truth. But we can say, just as exactly, that the experimental truth is that the value of the ratio is $\sqrt{2}$; the question which arises, from the empirical point of view which we are taking, is this: is the value $\sqrt{2}$ more or less easy to remember and to handle than the value 1·414213?

It may be worth observing that if for a physical constant (density, index of refraction, &c.) measurement gave the value 1·414213, it would hardly occur to the physicist to propose $\sqrt{2}$ as the value of this constant. Why? Simply because in a table of numerical values of analogous constants, all calculated in decimal form, there would be more inconveniences than advantages in writing in irrational form *one only* of the numbers in question. If on the contrary measurements of a certain class of physical constants furnished numerical results capable of being expressed, to the order of approximation of the measurements, by $\sqrt{2}$, $\sqrt{3}$, $\sqrt{5}$, $\sqrt{7}$, $\sqrt{13}$, $\sqrt{19}$, there is no doubt that there would be some practical advantage in adopting this representation, apart from the theoretical advantages which possibly might follow.

4. The Mathematical Continuum.

From the general conception of irrational number, we deduce that of the mathematical continuum, that is to say, of the aggregate of all real numbers, rational or not. This aggregate possesses, like the aggregate of rational numbers and, if we may so speak, *a fortiori*, the property of being everywhere dense; but it differs from that aggregate in two essential characteristics; (1) it is *perfect*; (2) it is *not countable*. Let us recall briefly the nature of these two properties.

Consider the series of values, approximate in defect, for $\sqrt{2}$, within $\frac{1}{10}$, $\frac{1}{100}$, $\frac{1}{1000}$ ...;

$$1\cdot4;\ 1\cdot41;\ 1\cdot414;\ \ldots$$

this is an increasing sequence of rational numbers; when we consider the rational numbers as the only numbers which exist, this sequence does not tend to a limit, for its limit is the irrational number $\sqrt{2}$, this being the way in which, from the purely arithmetical point of view, we define that number. We say that the aggregate of rational numbers is not *perfect*, because it does not contain all its limiting points; on the contrary, the aggregate of all numbers, irrational as well as rational, is *perfect*, for it is identical with the aggregate of its limiting points. This property is essential in every classical demonstration which involves the notion of continuity, so important in analysis; that is why the introduction of irrational numbers is indispensable for the construction of analysis.

But this positive advantage carries with it one considerable difficulty; the aggregate of the elements (geometrical points or numbers) which constitute the continuum is not countable, or, in other words, it is not possible to arrange these elements in a simple sequence the successive terms of which can be *numbered* by means of the natural sequence of the positive integers. I shall not develop here the reasons for which this character of not being countable appears to me to be a negative character, if we take it in its usual classical form.[1] I should like merely to remark that the aggregate of irrational numbers, effectively defined and used, is necessarily countable; for every possible class is countable (for example, the aggregate of algebraic numbers is countable) and the number of classes defined and in use by mathematicians at a given moment is limited. It is this aggregate of numbers effectively used which constitutes the practical continuum of mathematicians, the only one which they really use. The non-countable theoretical continuum is a metaphysical conception with which we have nothing to do here.

In this restricted aggregate, since enumerability reappears, we are able to define the concept of simplicity by an analogous procedure to the one which was used in the case of rational numbers. A rational number was regarded as the simpler the more briefly it could be defined in terms of unity. In the same way, *an irrational number will be regarded as the simpler the fewer the words required to define it* in terms of unity or, what amounts to the same thing, in terms of rational numbers.

To this criterion of simplicity, we ought to add in certain cases a criterion of *importance*, which to a certain extent overlaps the other. It sometimes happens that a certain

[1] I have indicated these reasons in a communication made to the Fourth International Congress of Mathematicians (Rome, 1908), and developed them in a note: " Les ' paradoxes ' de la théorie des ensembles " (*Annales de l'École Normale*, Oct., 1908). In the latter note I point out how, in my view, all aggregates effectively considered being countable, the important distinction from a practical point of view is the following: some are *effectively enumerable*, and the rest are not. I say that an aggregate is effectively enumerable when we can actually indicate a means of assigning a definite place to each of its elements, without possibility of ambiguity. See also the second edition of my lessons on the theory of functions (Gauthier-Villars, 1914).

irrational number occurs in a very great number of definitions; in other words, a very great number of definitions, apparently distinct, lead to the same number. Such is the case notably with the two celebrated numbers which analysts have come habitually to denote by the letters e and π. For these numbers, even if any one definition is more complicated than the definition of $\sqrt{43}$, for example, the fact that there exist so many definitions, relatively simple and connected with each other in numerous ways, leads us, on account of their importance, to classify these numbers as among the simplest. When the result of a calculation is the number e, or the number π, or a simple function of these numbers, everyone would agree to call the result a simple one.

We can summarize the above remarks by saying that the practical mathematical continuum includes, besides rational numbers, those irrational numbers whose definition is simple. This *simplicity*, moreover, is partly relative to the nature of the question under consideration, but *complexity* (or lack of simplicity) is necessarily limited by the nature of things; we have never—and for good reason—used a number whose definition, to be complete, requires a million volumes of 500 pages each.

5. The Practical Value of the Continuum.

We can understand now, without dwelling on the matter, what great practical value there may be in numbers which turn up naturally in mathematics in the course of the study of abstract problems. Take for example the number π. The analyst can give many definitions of it; he can say, for example, that the general integral of the differential equation

$$\frac{d^2y}{dx^2} + y = 0$$

admits the period 2π. This definition allows us to calculate

π to as high a degree of approximation as we desire; we find thus the earlier decimals to be

$$\pi = 3{\cdot}1415926535\ldots.$$

Suppose now, forgetting historical fact for a moment, that the above calculation had preceded the approximate evaluation of the ratio of the circumference of a circle to its diameter; empirical measurements, or a rough study of the simplest regular polygons, lead to an approximate value of that ratio which coincides, to a certain number of decimal places, with the value just written down; the coincidence could not be perfect, since every empirical value is necessarily inexact; but it might be near enough to suggest the idea of a closer relationship. Ought we then to conclude at once that the ratio of circumference to diameter is the number π previously defined by the differential equation? The induction would be rash; to justify it, a suitable course would be to bring into connexion the two notions, apparently quite distinct, of the differential equation and the circle; we should find that the integral of the equation is expressed precisely by means of the circular functions, *sine* and *cosine*, studied in elementary trigonometry; the values of these functions are given by ordinary tables of logarithms, and are used in a multitude of calculations; the relationship suspected is thus established. It may be added that this is not a mere matter of theory; the relations involved are of constant use in all practical questions in which trigonometry plays a part; in particular in the calculations of astronomy and geodesy, which depend on measurements of extreme precision; in the calculations of navigation also, less exact than the others mentioned, but of more obvious importance for daily life.

Such are the profound reasons which cause us to look upon the number π as not merely a *jeu d'esprit*, fitted to satisfy the taste for exactness and elegance of a few thousand persons who have made a deep study of mathematics; even from the purely practical point of view its true value (its exact value, in the absolute sense of the phrase) is *more exact*, that is to say, more useful than the approximate value written down above, even though direct measurement of the circumference and diameter of a circle could not possibly disclose the inexactness of the approximate value.

Many minds are haunted by this question of the value of π or the quadrature of the circle; some become unbalanced, and it happens quite often that men of sound mind on every other question publish papers to prove that the true value of π is not the " official " value; some propose $\sqrt{10}$, others $\sqrt{3}+\sqrt{2}$. The latter value is relatively closely approximate; we have

$$\sqrt{3}+\sqrt{2} = 3{\cdot}146\ldots,$$

so that the error is about 0·005; this is difficult to detect experimentally, seeing that it corresponds to 5 millimetres upon the circumference of a wheel 1 metre in diameter; an error of 1 millimetre in the measurement of the diameter would lead to a difference almost as great. But, even if this approximation were sufficient for the needs of practice, it would not be correct to say that the number $\sqrt{3}+\sqrt{2}$ is practically the exact value of the ratio of circumference to diameter, for we should thus be led in many questions to results which could not be accepted. It would not be quite the same if an expression as simple as the preceding gave the value of π correctly to ten places of decimals; such an expression would have a value by no means negligible; and it is probable that it would also be of considerable theoretical interest; but no expression of this kind has been found, nor, for reasons to be pointed out presently, is one likely to be found.

One final remark about the number π; certain patient calculators have found its value to 700 places; what interest has a calculation like that? Evidently none, so far as practice is concerned; but from the point of theory it may have very great interest indeed, if it suggests a simple law which we might possibly be able to prove; if, for example, we found that the figure 7 is always followed by the figure 9, or that the even figures are decidedly more numerous than the odd; but the existence of such laws is scarcely likely, and a result of a different sort such as this, " the 500th decimal of π is an 8 ", while true in the abstract, has no practical value, because it does not correspond to anything real; it is related to nothing else; we might compare it to the minutely exact determination, by processes of the greatest precision, of the weight of a single grain of corn chosen at random from the crop of a field; the number so obtained evidently represents a certain fact, but it has no scientific interest; it would be a different matter if we determined the mean weight of all the grains of corn, and the law of deviation of individual weights from the mean.

6. **Numerical Approximations.**—We have already recalled the fact that the theory of continued fractions presents us with a regular procedure for obtaining the simplest rational numbers approximating to a given irrational or experimental value; similarly we may try to find out the simplest expressions of any kind which give an approximate representation of a number given by experience. I shall not dwell on this question, which is a very difficult one, and has been little studied; I should like, however, to say a word or two about one feature which presents itself at the outset of the study, and which appears to be a general law.[1] It may be stated in this way: it is not possible to approximate to simple numbers by other simple numbers. An example will elucidate the meaning of this rather vague general statement. We may try to find an ordinary fraction equal to the square root of a given integer: to $\sqrt{2}$, for example, or to $\sqrt{43}$. If we denote the required fraction

[1] See, for the development of this idea, É. Borel, " Contribution à l'étude arithmétique du nombre *e* ", *Comptes rendus de l'Académie des Sciences de Paris,* 6th March, 1899, Vol. 128; and *Leçons sur la théorie de la croissance* (Gauthier-Villars, 1909).

by $\frac{p}{q}$, we can prove easily that the error made is about $\frac{1}{q^2}$, except for a factor the value of which has a limit depending on the data; for example, if it is $\sqrt{2}$ we are dealing with, the error is certainly greater than $\frac{1}{3q^2}$. We cannot therefore have an approximate value within one ten-millionth, for example, unless the terms p and q of the fraction exceed 1000.

On the contrary, it is evident that complicated numbers can sometimes be approximated to very closely by very simple numbers: thus the number

$$\frac{2}{3} + \frac{1}{10^{20}} + \frac{2}{10^{200}} + \frac{1}{10^{20000}} + \ldots$$

is obviously equal to $\frac{2}{3}$, with an error which affects only the twentieth decimal place. It was Liouville who, in his researches on algebraic numbers, was the first to give examples of the preceding law, the general proof of which would call for further research, but which appears really to be a general property of numbers. This is the *practical* justification of the introduction of numbers with simple definitions, other than the rational numbers: if, by replacing $\sqrt{2}$ by $\frac{3}{2}$ or $\frac{7}{5}$ we make an error which is practically always negligible, this substitution will be all to the good, and the exact value will retain only a purely abstract and theoretical interest.

7. **The Physical Continuum.**—M. Henri Poincaré gives the following " formula " of the physical continuum, as experience reveals it to us:

" We have observed, for example, that a weight A of 10 grammes and a weight B of 11 grammes produce in us identical sensations, and that, similarly, the weight B cannot be distinguished from a weight C of 12 grammes, but that we can easily distinguish the weight A from the weight C.

The rough results of experiment can therefore be expressed by the following relations:

$$A = B$$
$$B = C$$
$$A < C,$$

which can be regarded as the formula of the physical continuum."

It does not appear to me possible to subscribe to the theory proposed by M. Poincaré; the schematic experience that he imagines only fails to lead to a contradiction in consequence of its imperfections; a complete and total experience cannot contradict the principle of contradiction. By this I do not mean that we are to increase the precision of the experiments considered by M. Poincaré; to do so would only push the difficulty back a step, the only advantage of which would be to enable us to deny the possibility of actually carrying out the experiments described; but that is an argument which probably would not satisfy M. Poincaré. What I mean to say is that two magnitudes A and B ought not to be regarded as *empirically equal*, at a given period, unless it is impossible to detect a difference between them by any experimental means available at that period. This definition cannot lead to any contradiction, because if a careful experiment proves that A is less than C, while B, under the same conditions, cannot be distinguished from C, we have a sufficient experimental reason for asserting that A is less than B.

If we measure a magnitude by comparing it with a fixed standard, we obtain so many decimals certainly correct followed by a final decimal about which there is some uncertainty; different experimenters, equally careful, obtain different values for this last decimal, so that the observations can be summed up in some such form as the following: 30 per cent of the observers find a 5, while 50 per cent find a 6 and 20 per cent find a 7; another magnitude very near the first leads to results which are analogous, but not in general identical. The phenomenon is analogous to that which takes place in the *sophism of the stack of corn*.[1]

With reference to the passage quoted, an observation of quite a different kind may be made, and will be found in fact in M. Poincaré's book; his argument demands, not the mathematical continuum, but only a dense aggregate included within the continuum, for example the aggregate of rational numbers; because, however near two rational numbers may be, we can divide indefinitely the interval between them.

Let us be content, then, to state the following truth, which may appear to be evident *a priori*: the physical continuum is distinguished from the mathematical continuum by the fact that experiment never permits us to attain more than a limited degree of approximation, so that a certain minimum difference is necessary before we can distinguish between elements which are nearly equal.[2]

8. **The Relations between the Two Continua.**—We may infer from the preceding remark that, for the purposes of the physical continuum, we need only consider decimals with a fixed number of figures, this number varying according to the nature of the subject. Thus has arisen the idea of a *mathematics to five decimals*. The chief objection to this conception is that it is not invariant, even with respect to the simplest operations; but, without waiting to develop

[1] See É. Borel, " Un paradoxe économique: le sophisme du tas de blé ", *Revue du Mois*, 10th Nov., 1907, Vol. IV, p. 688; and *Le Hasard* (Alcan, 1914).

[2] With respect to this statement it does not follow that M. Poincaré's objection is still fatal; the minimum difference is not an absolute constant, but depends on experimental conditions and often also on the value or the nature of the quantities we are measuring. Consequently, two magnitudes A and B, which direct processes of measurement do not enable us to distinguish, can be differentiated if we have the happy thought—or the chance—of comparing them both with a third suitable magnitude C, which direct experiment does not separate from B, but does from A; these new experiments diminish the *minimum separabile* for A and B. Another method for diminishing this minimum is repetition of experiments and application of the calculus of probabilities; this is a very interesting question to which I hope to return on another occasion.

this objection, we have sufficient reasons of a positive kind for rejecting *mathematics to five, or to seven decimals.* We have seen, in fact, that in many cases the true solution of a practical question, the true value of a ratio to be measured, is a number, rational or irrational, which cannot be expressed exactly in the form of a decimal fraction, but the definition of which is simpler than that of the decimal fraction which represents it to the degree of accuracy of experiment.

We thus find ourselves returning, as it were, to a tendency which in principle is as old as Pythagoras: to try to express all magnitudes in a simple way by means of whole numbers. Only, it appears to us legitimate to add to the rational operations which we can effect on these numbers the simplest irrational or transcendent operations.

We know that in many experimental questions a simple expression of the kind we are discussing has not been found and probably does not exist: we ought then to content ourselves with mathematics to five or seven decimals, the arithmetical construction of the continuum being useless. In other questions, on the contrary, known laws enable us to put in evidence experimentally simple rational numbers, the square roots of whole numbers, the number π, &c. There is, therefore, an intimate connexion between the mathematical continuum and the physical continuum; the arithmetical notion of number really plays a part in the study of nature, and not merely the empirical notion of approximate measurement. This is still true, moreover, in those cases, apparently numerous, notably in the theory of atomic weights and in crystallography, where the simple arithmetical value plays solely the part of a first approximation.

NOTE IV

The Universe—is it Infinite?

1. A Finite Universe is Possible.—At the end of one of the lectures which he gave at the Collège de France in March, 1922, M. Albert Einstein pointed out briefly the reasons for which the general theory of relativity led us to think that the universe is finite. Among the consequences of the new theory this is evidently the one which is the most difficult to verify and the least important for practical purposes; but it is perhaps the most interesting from the point of view of philosophy; is not the question whether space is finite or infinite one of those which have most intensely interested philosophers, ever since there were philosophers? The solution given by Kant to the antinomy of finite or infinite space appears to many minds as an evasion rather than a solution; in any case, on its scientific side, the only one which I can venture to take if I would not risk stumbling at the very first step, the problem still awaits solution.

It may perhaps be objected that it is impossible to form for ourselves any definite image of infinity by the light of science alone; that consequently, whether we would or not, we cannot escape from the metaphysical side of the problem. To that one might answer that even if we admit provisionally that an infinite world goes beyond the limits of positive science, still we cannot deny that a finite world is a perfectly clear scientific conception; if then we have reasons for believing that this conception corresponds to reality, the hypothesis of an infinite universe is by that very fact at once ruled out.

A first objection to the finite world is drawn from the fact that a finite body appears to us as bound to have limits; and we cannot but ask in that case why the limit should be impassable. This objection loses its force in the light derived from study of modern geometries, as we have seen in Chapter IV.

Nevertheless, the fact must be insisted on that these considerations of analytical geometry are pure abstractions, which do not enable us to reach reality; all that they give us, and it is certainly much, is the assurance that a finite universe is possible. Is it real? That is a physical question which only observation and experiment can give us the means of trying to answer.

2. **The Mean Density and the Curvature of the Universe.**—The new light which Einstein's theory casts on the subject is this: it establishes a connexion between the density of matter and the curvature of the universe. Let us try to explain the nature of this conception of curvature of the universe. In a plane, we say that a straight line has no curvature, that is to say, a curvature equal to zero, while a circle has a curvature which is the greater the smaller its radius. For definiteness, we shall take for the value of the curvature the reciprocal of the radius, measured with a unit of length chosen in advance. If this unit is 1 metre, a circle of radius 1 metre will have the curvature *one*, while a circle of radius 100 metres will have the curvature one-hundredth, and a circle of radius 1000 kilometres will have the curvature one-millionth. In the last case, the curvature is so slight that we should have difficulty in distinguishing the arc of such a circle from a straight line; there is, however, this essential difference that the straight line extends to infinity, while an arc of a circle of 1000 kilometres radius, if we prolong it indefinitely with the same curvature, will end by closing in on itself; it is finite, while the line is infinite.

The definition of curvature for surfaces is more complicated than for curves; it is still more complicated for solids in three dimensions; but the essential property remains; if the curvature is greater than a certain number, no matter how small this number is, the surface or the solid cannot exceed a certain definite total size, which is simply related to the reciprocal of the value of the curvature.

But, in Einstein's theory, the curvature of the universe at a point is simply related to the density of matter in the neighbourhood of the point. We conclude that, *if the mean density of matter is greater than a certain fixed number, no* matter how small this number is, the universe is necessarily finite and, consequently, the total quantity of matter is itself finite.

It will help to make the import of this conclusion thoroughly clear if we explain a little more definitely the meaning of the hypothesis which is made with respect to the mean density of matter.

This mean density is extremely minute, in accordance with the emptiness of interstellar space; the star nearest the sun is so far away that its light takes three years to reach us; the mass of matter contained in a sphere whose radius is measured by three light-years is therefore not greater than the mass of the sun,[1] that is to say, the mass of a sphere whose radius is traversed by light in little more than two seconds. But three years are equivalent to more than 80 million seconds; the ratio of the radii of the two spheres being 40 millions, the ratio of their volumes is the cube of 40 millions; or more than 60,000 milliards of milliards; this is the number by which we must divide the density of the sun to get the mean density of the matter in the greatest sphere which has the sun for centre and which does not contain any other star. But, small as this number is, if the mean density does not become smaller when we take spheres of greater and greater radius, the conclusion is definite: the universe is finite. For this conclusion to fail, it is necessary that the mean density should become smaller and smaller as the radius of the sphere increases. The universe

[1] The mass of the planets is a very small fraction (a thousandth) of the mass of the sun; it is negligible here.

could then be compared to a curve like a hyperbola or a parabola, the curvature of which is always in the same sense, but which nevertheless extends to infinity, because this curvature becomes smaller and smaller as we go farther and farther away. The arc of a hyperbola has an asymptote, the parabola has not; we need not trouble about this difference here.

3. **The Hypothesis of an Infinitely Small Mean Density.**—The hypothesis of a mean density of matter tending towards zero for a sphere of continually increasing radius may be described more briefly as the hypothesis of the infinitely small mean density of the matter in the universe; it is not Einstein's theory alone which leads to this hypothesis as a necessary condition for the possibility of an infinite universe.

It has been known in fact for many years that if we assume that light is not absorbed in interstellar space (and this hypothesis appears to correspond closely to the facts), then a distribution of the stars with any approach to regularity would entail the consequence that the brightness of the whole sky would be comparable to the brightness of the sun. To see how this comes about, imagine that we found ourselves in a very sparsely sown forest; no matter how sparsely sown it is, if it is of sufficient size, the horizon will be completely hidden from us by the trunks of the trees. In the same way, no matter how widely scattered the stars are, if they extend over a universe which is infinite, and do not thin out more and more as we go farther away, there will not be a single piece of the sky, no matter how minute, which is not covered by stars. Under a magnification much stronger than that of our telescopes, these stars would appear like circles overlapping each other and leaving no space unoccupied: but the brightness of a luminous surface does not depend on the distance when there is no absorption; the brightness of the sky will therefore be all over of the same order as the brightness of the sun. It is therefore necessary, to account for the appearance of the starry sky, to suppose that the universe is built up of matter in the following way:[1] the stars which we see practically all belong to the system of the Milky Way; they constitute what we may call a galaxy of the first order. A very large number of galaxies of the first order at immense distances from one another will constitute a galaxy of the second order; similarly, a very large number of galaxies of the second order, at still greater distances, would form a galaxy of the third order, and so on; we can imagine that the successive distances increase fast enough for the mean density to become smaller and smaller as the order of the galaxy we are considering becomes higher. We thus arrive at a conception of an infinite universe which is not inconsistent with Einstein's theory, or at least with the more elementary consequences of that theory.

To go further, it would be necessary to examine more closely what, in Einstein's theory, happens in the case of those parts which are extremely remote in a universe of infinitely small mean density; we should then encounter complications of such a nature that we may be tempted to discard them as highly improbable. If we accept the metaphysical postulate that the world ought to be intelligible to us without it being necessary to introduce excessive complications, we may thus be led to grant to Einstein that it is *convenient* to suppose that the universe is finite, in the sense used by Poincaré when he said that it is convenient to assume that the earth turns round the sun. For Poincaré, this criterion of convenience becomes the test of scientific truth, as soon as we assume that science is possible, for the

[1] The detailed calculations, which we omit, have been developed completely by one of the most learned investigators in stellar astronomy, M. C.-V.-L. Charlier, director of Lund Observatory (Sweden), in his recent paper: *How an Infinite World may be Built Up*. See, on the other hand, my note: "Définition arithmétique d'une distribution de masses s'étendant à l'infini et quasi périodique, avec une densité moyenne nulle" (*Comptes rendus de l'Académie des Sciences*, Vol. 174, p. 977, 10th April, 1922).

complication introduced into mechanics by the geocentric hypothesis of Ptolemy would make all science impossible. If, after verification, the hypothesis of an infinite universe, combined with Einstein's theory of gravitation, lands us in difficulties comparable with those of the geocentric hypothesis, that may be for many minds a sufficient reason for abandoning it.

4. **Of what Use are these Cosmological Speculations?**—These bold speculations, as we can now see, are at once of surpassing interest, and, it must be carefully recognized, lacking in that rigour which is to be found in other parts of science. It may seem rash indeed to draw conclusions valid for the whole universe from what we can see from the small corner to which we are confined. Who knows that the whole visible universe is not like a drop of water at the surface of the earth? Inhabitants of that drop of water, as small relative to it as we are relative to the Milky Way, could not possibly imagine that beside the drop of water there might be a piece of iron or a living tissue, in which the properties of matter are entirely different.

These reservations were necessary, but do not appear to me of such a nature as to destroy all interest in these cosmogonic speculations. As Boltzmann wrote with reference to considerations of the same order:[1] " Assuredly, no one will take such speculations for important discoveries, nor for the highest aim of science, as did the ancient philosophers. But it is not at all certain that it is right to laugh at them and to treat them as altogether useless. Who knows that they do not enlarge the horizon of our circle of ideas, and even contribute to the advance of experimental knowledge, by increasing the mobility of our thoughts?"

[1] *Leçons sur la théorie des gaz* (translation by Gallotti and Bénard), Vol. II, p. 252.

INDEX OF NAMES

GENERAL INDEX